HARRY MARSHALL WARD
and the
Fungal Thread of Death

By Peter Ayres

Lancaster University
United Kingdom

APS PRESS

The American Phytopathological Society
St. Paul, Minnesota

Cover: Photo of Harry Ward Marshall by permission of the University of Cambridge, U.K.; drawings of wheat rust fungus, *Puccinia graminis*, by H. M. Ward, *Annals of Botany* 2:217-222, 1888.

This book has been reproduced directly from computer-generated copy submitted in final form to APS Press by the author of the volume. No editing or proofreading has been done by the Press.

Library of Congress Control Number: 2005923941
International Standard Book Number: 0-89054-333-X

Printed in the United States of America on acid-free paper

The American Phytopathological Society
3340 Pilot Knob Road
St. Paul, Minnesota 55121, U.S.A.

Contents

Prologue

It is remarkable that Harry Marshall Ward reached the pinnacle of his profession, became a Fellow of the Royal Society of London, was Professor of Botany in the University of Cambridge, and guided a King. Although he was born in 1854 into a family which prized education, the arts and learning, his parents struggled financially. They belonged to the genteel, lower middle-classes. And class was important because, in mid-Victorian Britain, few succeeded in moving upwards, crossing boundaries marked by differences in wealth and education, in a society where reputation and personal contacts counted heavily. Harry had a fine intellect and, as his later life demonstrated, an exceptional capacity to work passionately and single-mindedly towards a goal. However, his prospects were limited because his father was a poorly paid music teacher who, when Harry was a teenager, contrived to lose what little capital the family possessed. Neither his family nor their friends were in a position to help the young man. Whatever he was to achieve in life he would have to achieve through his own efforts.

Luck was on young Harry's side, however, because by the time he had reached his teens the nation had recognised that, for its continued prosperity, it needed more schoolteachers and it needed more science to be taught. A scheme had recently been introduced which took the brightest students, irrespective of background, and trained them at South Kensington, in London, to become schoolteachers of science. He spotted his opportunity, studied in whatever time he had spare from working to supplement the family income, and won a coveted training place.

At first it was enough for him to think that one day he would be able to teach science, especially his first love, biology, but as he settled into his new life in London he realised that he wanted more. He wanted to take a degree and to become a researcher. And then good luck intervened for a second time, turning his dream into a reality. He had made friends with Louis Lucas, a young man of his own age who occasionally dropped into the classes at South Kensington. Lucas was interested in biology but, coming from an extremely rich family, he had no need of a teaching qualification. He did, however, recognise Harry's outstanding ability as a biologist and, believing that such a talent should not be wasted, he guaranteed to finance Harry's studies at university, if he could pass the entrance exams. Harry did, of course, pass the exam and went on to earn a first class degree from the University of Cambridge. It was the beginning of a brilliant career.

The structure of the universities – what was taught and by whom – mirrored, indeed it helped to underpin, the unchanging, rigid society of the mid-Victorian age. The way botany was taught was a paradigm for the sciences. Little new knowledge was generated and it was held that that which was discovered should be allowed to stand the test of time before being introduced to students. Whereas observations of the anatomy and familial relationships of plants were begun by the Ancient Greeks, and continued in the following centuries, *experiments* on plants were very rare. Textbooks went unchallenged and hypotheses went untested. In continental Europe the position was much the same but, by 1850, the first signs of change were detectable in French and, more particularly, German universities. In the 'New Botany' that developed, students were encouraged to challenge accepted wisdom: research was prized and students were taught to rely not on textbooks but upon their own observations and experiments. Harry and many of his generation of young botanists determined to visit Germany to study with the leading professors, Julius Sachs in Würtzburg and Anton De Bary in Strassburg. Inspired by his experiences in Germany, Harry decided to devote his life to what he and his friends called 'The Cause', the establishment of a vigorous botanical school in Britain, independent of Germany.

Harry was The Cause's most passionate disciple – his friends thought too passionate, for over-work may have contributed to his early death – but after much frustration and many setbacks he eventually achieved his goal. As professor of botany in the University of Cambridge, and through the new teaching and research that he fostered, he established there the new botany that was so dear to him. The intellectual energy that had fired the German universities was slowly extinguished by stifling political changes. Leadership of the new botany passed to Britain where Harry was tireless in his support of those young scientists and new disciplines that would carry it forward. For a generation after his early death, the Cambridge Botany School was at the forefront of the world of plant sciences.

At branching points in the history of a science we often find extraordinary men. In the history of plant pathology, Harry was among the first to study the physiology and biochemistry of the struggle between host and pathogen. He was a pioneer in emphasising the effect of the environment on the outcome of that struggle. By example, he helped to found a whole new sub-discipline, physiological plant pathology, which has sought to explain in physical and chemical terms the interactions between plants, their pathogens and the environment. In retrospect we can see that, in attempting to explain the mechanisms whereby infections can be prevented or halted, the subject has occupied the middle ground of plant pathology and has dominated the parent discipline.

Harry Marshall Ward's reputation has until now rested on discoveries about the transmission of plant disease that he made while studying coffee leaf disease in Ceylon, during his first employment after graduation. Important as these were, both biologically and in establishing his reputation as a researcher, historical perspective shows that they are much less significant than his role in establishing the pre-eminence of British botany in the early years of the 20th century and his part in the origins of physiological plant pathology. Neither of these has been properly recognised before and they will form the core of this biography.

The book traces Harry's development both as a scientist and, in order that the scale of his achievements can be appreciated, as a man of his times. It begins with his time in Ceylon, for this was a period when, living far away from everything that was familiar, he was forced to come to terms with his own ambitions and abilities. This process, sometimes painful, is revealed in a series of letters he wrote to his family at home. What emerged from these formative years in Ceylon was a man able to build a career of the highest distinction, a scientist whose rich legacies are our inheritance.

Chapter one: To Ceylon for coffee

While travelling from coffee plantation to coffee plantation *"I often come upon rough fellows living in the merest mud hovels, sleeping on a few planks, eating and drinking of the meanest, and yet I find this rough exterior covers the soul and intellect of an educated gentleman. Such an one was here. ... He cannot come home for want of money, and his coffee costs more than it yields: he goes about the barren acres in a moody and despondent way that speaks of disappointed hopes, misplaced talent and wretched despair.*

In one dull corner of that dark, broken-windowed bungalow I discovered a well thumbed volume of Beethoven's Sonatas lying on a very small, very plain, and very cheap harmonium: after dinner the man made me wonder and think of many things as his huge rough brown hands played some of the dearest and sweetest music. ... I would defy the world to find a better place for true music or a better feeling on the part of the player. He had chiefly taught himself in his lonely hours far away in those gloomy forests and hills. When he learned that I too liked music without criticism, he was a changed man, and we sang and played late into the night: upon my word I never felt nearer loving a man in my life than when we sang that beautiful duet – 'I would that my love in one found words' ... in the midst of the dark, wild jungle."

Harry Marshall Ward writing from Peradeniya, Ceylon, 20th April 1881, to his mother and family in Nottingham, England.[01]

The Government Cryptogamist arrived in Ceylon (Sri Lanka) early in 1880. The slim upright man young man, with a mane of dark hair and a bushy moustache, who descended the gangplank from the MMS 'L'Ava' in Colombo harbour was Harry Marshall Ward (Fig. 1:1). The investigations of Coffee Leaf Disease that he would carry out in the next two years would bring him the highest reputation with his fellow botanists and would set in motion a career that, ultimately, would lead to a Fellowship of the Royal Society and the Chair of Botany at Cambridge University, his *alma mater*. But for the moment, he was just an apprehensive twenty-five-year-old, fresh to the tropics and all too aware of the high expectations of him held by the Colonial Government, and by the coffee planters it represented. He would have been conscious too that his personal status was changing as he stepped ashore in Ceylon. He had been well educated, but he was from a socially undistinguished and relatively poor middle-class family in Nottingham, England. Now, as a government official, he would be expected to display social skills befitting his position. He would dine with the Governor and later in the evening be invited to take his turn in the round of songs and recitals.

Fig. 1:1. Harry Marshall Ward photographed in Colombo sometime in 1880–82. (By permission, the Royal Botanic Gardens, Kew.)

He would not be well paid – indeed, financial struggles were always a feature of his life – but in his dress and decorum he would be required to maintain the high standards of the colonial ruling classes. On all sides, much was expected of him. It was therefore with considerable trepidation that the young man stepped into the world of the British Empire.

Throughout the previous two hundred years the Empire had grown to dominate the world. It had done so thanks largely to the efforts of countless thousands of Britons prepared to suffer extremes of hardship and loneliness in distant lands, sometimes purely for the sake of adventure but more often in the hope of gaining power and wealth. As soldiers, colonial administrators, and businessmen, these young men and women had transformed the countries swallowed one by one into their expanding Empire.

Plants were central to the whole enterprise of Empire-building. In the first phase, involving domination, there was the need for the colonists to secure a plentiful supply of food. In the next phase the emphasis turned to the growth of cash crops. Either, indigenous crops were grown on a scale vastly more extensive than ever dreamed of by the natives, or alien crops were introduced with scant regard to native sensitivities or, too often, to ecological principles. Starting with the development of sugar cane, tobacco and cotton crops in North America and the West Indies – and having well known consequences in the birth of the slave trade – the Empire had been instrumental in the development of tea estates in northern India, in the planting of tea, tobacco, cocoa, and sisal crops in Africa, and of wheat and sugar cane in Australia. Some crops were grown for medicinal purposes, such as cinchona (the source of quinine) in southern India, while others were grown for industrial purposes, for example, rubber in Malaya. Crops were often developed for strategic reasons, the object being to replace or suppress products from the competing Dutch, French or German Empires.

New plant species were collected at first simply for their intrinsic interest but, in the later years of Empire, they were increasingly recognised as potential sources of novelty for the gardens of Europe and North America. Small fortunes could be made by finding plants which captured the mood and imagination of Victorian gardeners. Thus, typically financed by the Royal Botanic Gardens, Kew, or by commercial seeds companies, 'plant hunting' was at its peak from the late 19th through to the early 20th century. Among those young adventurers who joined early expeditions were Joseph Banks and Joseph Hooker – both to become the most eminent of botanists – and later Harry's own son, Frank Kingdon-Ward, who has been called the last of the great plant hunters.

Modern British commentators[02,03] are able to argue that the builders of the British Empire were not entirely ruthless, but were imbued with at least some selfless ideals. An interest in botany was a mark of cultivated Britons abroad. By the end of the 19th century, letters, journals and memoirs were filled with botanical information; correspondents drew and pressed flowers and other plants; and travellers regarded the ubiquitous botanical garden as an important sight on their tours. In Calcutta, Cape Town, the West Indies and many other centres of Empire, Botanic Gardens were founded that are still world famous today. Some of these gardens were directly linked to Kew in an extensive imperial research network. Sometimes, like Kew, they served as international quarantine stations when potential crop plants were moved around the world. In Ceylon, the Royal Botanic Gardens at Peradeniya were established in 1821, four miles from the old capital, Kandy. The greater part of the land had been a royal demesne occasionally occupied by the Kings of Kandy and was probably once a fruit garden (Pera = guava; deniya = enclosed space). At roughly 500m above sea level, Peradeniya suited a wider range of plants than the first Garden at Slave Island, Colombo, founded in 1810 on the suggestion of Sir Joseph Banks, then President of the Royal Society. Peradeniya was to be Harry's home and workbase throughout his time in Ceylon.

Some time before 1000AD, Arab traders brought seeds or plants of coffee (*Coffea arabica*) from the Ethiopian highlands, where it grows wild, to the Yemen, where for the next 500 years the plant was cultivated but fiercely guarded from the outside, non-Islamic world. At some point in the 17th century, seeds were captured by Europeans and by the late 18th century Dutch planters were growing coffee crops on a relatively small scale in Ceylon and on several islands of the East Indies, including Java. However, after taking over Ceylon (1797) from the Dutch in the Napoleonic Wars, and particularly after 1825, the British expanded enormously the area of the island planted with coffee. They stripped out the mixed native vegetation from large tracts of the land in the higher regions, replacing it with coffee estates. Under the Governorship of Sir Edward Barnes, colonial policy changed from building forts to building roads that opened-up the interior of the island. Barnes himself took up coffee planting. The Director of the RBG Peradeniya was instructed to give his attention to "the growth of coffee which His Excellency has particularly at heart to see increased throughout the Island".[04] The mountain ranges on all sides of Kandy were rapidly covered with plantations. It is said that the East India Company's officers crowded to Ceylon to invest their savings, and capitalists from England arrived by every packet boat. The prospects were dazzling, "expenditure was unlimited; and in its profusion was equalled only by the ignorance and inexperience of those to whom it was entrusted".[05]

As with world commodities today, there were ups and downs in prices and demand. Fortunes were both made and lost but, by 1870, the island was the world's largest producer of what had become a hugely fashionable drink worldwide. It exported 45,000 tonnes per year and the crop covered 160,000 hectares of the central uplands. In 1867 a railway was completed linking Kandy with Colombo and its port, 75 miles away. Planters were no longer dependent on native-owned bullock carts to take the crop down to the coast and to bring back supplies. Just as importantly, their coffee exports were fresher and of higher quality. Money continued to pour in to the island and a modern infrastructure of roads and railways, banks and hotels continued to develop. On the plantations there was even a shortage of labour, which was solved by bringing Tamil people from southern India, less than 100 miles away across the narrow Palk Strait. Incidentally, the arrival of these Hindus solved a short-term problem but, by ignoring the cultural and religious differences between the Tamils and the native Buddhist Singhalese, the colonial government sowed the seeds of a bloody conflict which, 150 years later, nearly tore Sri Lanka apart.

There was optimism everywhere and no-one suspected the coffee trade was in fact at its zenith. Then, in the space of just a few years, it all went wrong. In 1869 a new disease had been detected on a single estate in the Mandulsima district. The first signs were a heavier-than-normal shedding of leaves by the evergreen coffee bushes in the 'autumn' season. Coffee berries maturing in the same season were not affected and it was easy to overlook the appearance of the whole bush because a new flush of leaves was soon

Fig.1:2. Visible symptoms of coffee leaf rust. All cultivars of *Coffea arabica* grown in Ceylon were susceptible, as was the recently introduced Liberian coffee, *C. liberica,* and the native *C. travancorensis*. (By permission of Professor James Baker.)

produced. It was only as the disease progressed, and the excess of leaf-loss over leaf-gain had significantly denuded the bushes, that the problem became obvious. George Thwaites, the Director at Peradeniya and senior botanist on the island, was alerted and became immediately concerned on seeing for himself the micro-symptoms of disease, the rusty-red spots on the leaves of infected bushes (Fig. 1:2). Indeed, he was so concerned by the threat he had recognised that he sent infected material back to England for an opinion from the Rev. Miles Berkeley, an elderly and deeply respected expert on fungi. (Although mycology, the study of fungi, was a hobby for Berkeley, this archetypal 'gentleman scientist' was one of the very few men of the 1840s who recognised that the Blight devastating potato crops in Ireland was caused by a parasitic fungus.) Berkeley, now 66 years old, realised that the fungus sent from Ceylon represented a new genus; he published a description of the fungus and called it *Hemileia vastatrix*. He recommended that the planters should try to protect their crops by the application of sulphurous solutions but his advice seems either not to have reached the ears of the planters or to have been ignored by them. To them, everything seemed normal and they did not recognise the signs of impending disaster. The planters were used to seasonal fluctuations in yield; their profits were up because the unit price of the crop rose by more than 50% in the mid 1870s, and any decreases in production from the older plantations were disguised by a vast extension of cultivation in lands around the hill town of Nuwara Eliya, up to 1,500m elevation. The underlying story is revealed however by quinquennial figures for exports showing that from a peak in 1871 they fell by 13.3% in the period 1876–80, and by 57.5% in 1881–85. By 1886, exports were down by 79.1%

and coffee growing was virtually finished in Ceylon. In Colombo the Oriental Bank closed its doors: most planters were ruined.

By 1879, 10 years after the leaf disease was first recognised, the planters were becoming painfully aware of the catastrophe overtaking them. Their strident demands for help forced the Colonial Government to set up a Commission of Inquiry, and officially sanctioned investigations were begun by Daniel Morris, Assistant Director at Peradeniya. Morris was destined for a long and distinguished career in botany but, as was betrayed by the inconclusive and vague report he soon produced, fungi were not his strong point. He was not a mycologist and, to make matters worse, it seems he was not personally liked by his boss, Thwaites. The investigations were soon suspended, Morris was sent to the Botanic Gardens in Jamaica, where his career flourished, and an appeal for help was sent to the parent Botanic Garden at Kew where the Director, Joseph Hooker, as so often, turned the matter over to his Assistant and future son-in-law, William Thiselton-Dyer. It was Dyer who had previously recommended Morris, but this time he got it right. He knew an exceptionally able young mycologist who just happened to be available – a certain Harry Marshall Ward. An appointment was quickly made. It has to be pointed out, Harry was at the time a specialist in the fungal diseases of plants only in theory. In reality, he was a mere 25 years old, had only just graduated from Cambridge University and had no practical experience of disease in the field. To make his situation worse, he was faced with an epidemic that had already gained momentum; an epidemic which could not have been stopped with modern chemicals. In his favour, there were few men in 1880 who did have any relevant experience and he was bursting with the ideas and lessons he had recently learned at Cambridge and in the Strassburg laboratory of the great Anton De Bary. This German professor's knowledge of plant diseases was far in advance of that of any of his contemporaries, and his new methodology, with its emphasis on detailed microscopic observations, complemented by experiments, was defining the nascent science of plant pathology.

The causal connection between microbes and infectious diseases, whether of plants, animals or man, seems so obvious and elementary to us today that it is hard to put ourselves into a world that was only just beginning to make the connection. To place Harry's task into context, it was only in 1861 that De Bary had proved *by experiment* that healthy plants (potatoes) become diseased (blighted) after inoculation with a parasitic fungus, while un-inoculated control plants remain healthy. Thus, just 19 years had elapsed since man had learned for the first time the lesson that epidemics among crops, like those which had caused famines since the beginnings of agriculture, are caused by microbial infection. The concept was still new and, like the parallel concept of the microbial origin of disease in humans, was not accepted by everyone.

Harry was faced not just with frustrated planters and government officials but with millions of dying plants whose leaves, in addition to the various damages caused by insect herbivory, bore every evidence of having been colonised not just by one fungus but by many. The cause of 'Coffee Leaf Disease', as it was called at the time, was far from obvious. Harry attacked the problem on two fronts. On one, he followed the methods of his German mentor. Using thin, hand-cut sections of diseased tissue he studied under his microscope leaves at different stages of infection by the putative pathogen. With painstaking observations, he methodically and exhaustively worked out the life cycle of *H. vastatrix*. On the other, he went beyond De Bary and did something totally novel. In order to find how a pathogen might spread from plant to plant, he trapped the reddish-brown spores, from which the fungus gets its common name, 'rust', on sticky microscope slides hung like leaves amongst the crop canopy. The difficulty of his task may be appreciated from a short extract taken from the Appendix to his Second Report[06] which describes what he found on one slide, previously boiled in sulphuric acid and washed in pure water, and exposed for 12h.

> On one portion of the upper surface, ¼ inch square, I recognised twenty eight individual spores, one mass of toruloid bodies, and a rather large fungus filament; in addition to these were larger or smaller pieces of amorphous organic matter, bits of cell wall, and inorganic particles. Crowds of brilliant particles (bacteria and microgonidia) and pieces of spider's web also occurred. In the twenty eight spores counted, seventeen kinds were noticed; one of these was a yellow papillate [warty] spore of *Hemileia*.

Whether or not Harry took his inspiration from Charles H Blackley,[07] a Manchester physician interested in hay fever, who some years earlier (1873) had trapped pollen grains and fungal spores on sticky slides at 'breathing' level and on kites flown at 300m, Harry's attempt to trap infective propagules from the air was totally new to botany. It was the first manifestation of his scientific greatness.

Later he was to modify his spore-collecting techniques. As he did so, he trapped more and more of the rust spores. By practical means he was able to demonstrate an epidemiological 'first'; he showed a plant disease can spread through aerially transmitted spores.

From microscopy, he found how each papillate uredospore of *Hemileia* germinates on the leaf surface, giving a fine thread, or hypha, which penetrates the leaf (Fig. 1:3). By cutting thin sections of leaves and tracing the progress of penetration and inter-cellular growth, he found how, after the thread-like hyphal branches ramified through the leaf, the fungus inserted "tapping" or "exhausting" organs (nowadays called haustoria) into the cells of the leaf; it next burst back to the surface, releasing into the air 150,000 or more spores from each small infection site. Given that each leaf might have tens of infection sites and the fungus can complete its life cycle in about 40 days, the air must sometimes have turned red on windy days – a terrifying sight! Most significantly, and after many failures, he learned to inoculate small areas of leaves on healthy plants with rust spores and to keep those plants under the warm moist conditions which induced them to develop, some 5–6 weeks later, characteristic rust lesions at the point of inoculation.

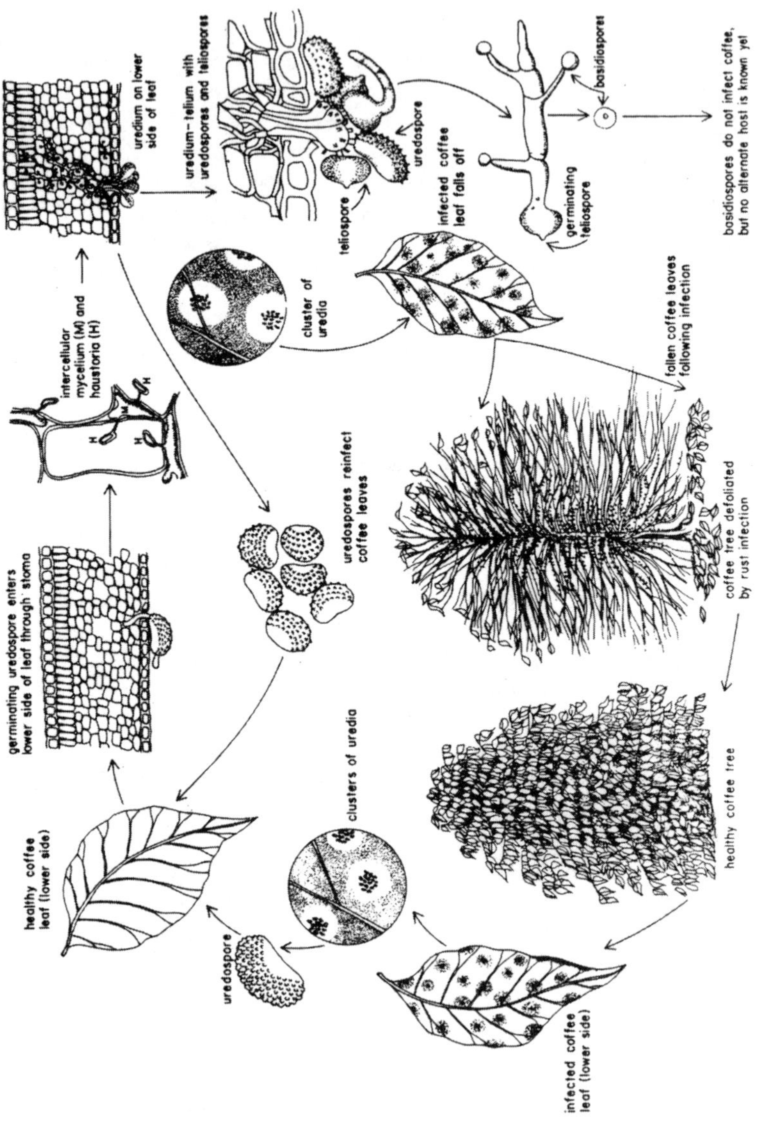

Fig. 1:3. The life cycle of *Hemileia vastatrix*. (Reprinted from Coffee rust, in: *Plant Pathology*, 4th ed., G. N. Agrios, ©1997, with permission from Elsevier.)

Harry had described every stage of an infection cycle, something which had been done for the first time with a plant pathogen only a few years before by Anton De Bary. Harry had proved Coffee Leaf Disease was caused by a parasitic fungus, *H. vastatrix*. He was not been able to culture the pathogen on an artificial medium, such as nutrient agar, but this was excusable because, as we now know, rust fungi are obligately biotrophic, i.e. they will only grow on a living host. Failure was due to the peculiar nature of the rust fungus and not to the scientist.

Rust fungi infect most families of flowering plants, each fungus species being specific to a different species of plant host. Among the crops affected by rusts are cereals and grasses (Gramineae), peas and beans (Leguminosae) and onions and leeks (Liliaceae), while trees from birch (*Betula*) to willow (*Salix*), and ornamentals from hollyhocks to roses are similarly afflicted. In being faced by the rust that affects coffee, Harry was actually very lucky for it has an extremely simple life cycle. Whereas some rusts have up to five different spore types in a complete life cycle, which may involve a second, unrelated plant host, *H. vastatrix* has just two spore stages, only one of which is functional, and no secondary host. He succeeded in identifying and distinguishing both spore types, the effective uredospore and the ineffective teliospore.

Harry was not so lucky where the availability of possible treatments for the rust disease was concerned. Field trials were carried out on several of the coffee estates. Carbolic acid and a range of chlorine compounds were considered but, realistically, the only fungicide available to Harry was a lime and sulphur mix. In practice, he confirmed it did not stick well to leaves and its control of spore germination was poor. The technology of 'formulation', i.e. the addition of agents to wet leaves and to aid retention of the chemical, had not yet been born. Lime-sulphur had no curative action. It was not a remedy against established infections. There was no cure that Harry could recommend to the growers.

It was not until 1885 that the world's first modern fungicide, Bordeaux Mixture, was announced by a Frenchman, Pierre Alexis Millardet. The Mixture was used originally to control mildew (*Plasmopara viticola*) on the leaves of grape vines but it is still widely used today by first-world gardeners and third-world farmers to control diseases such as potato blight. Millardet is remembered for recognising that

> "a practical treatment [of fungal disease] ought to have for its objective not the killing of the parasite in the leaves infected by it, which seems impossible without killing the leaves themselves, but of preventing its development by covering preventively, the surface of the leaves with various substances capable of making the spores lose their viability or, at least, of impeding their germination".[08]

Five years earlier, Harry had reached the same conclusion. He saw clearly that pathogenic fungi are vulnerable to the environment, including any fungicidal chemicals, while on the leaf surface; the first germination hypha that emerges from the spore wall is especially vulnerable until it gains entry into the relatively safe zone that is the nutrient-rich, moist environment of the leaf's cells and intercellular spaces. "The reagent must be on the spore at germination," he emphasised. Remarkably, his perceptiveness extended to an awareness of environmental risks. He warned, "the reagent must be soluble and yet when it is most required to be active is *when the rains are continuously dissolving it, and running off with it in solution to the ground* [author's italics]. This being so, one must be doubly careful in selecting nothing which (or products of which) may do harm to the tree on being washed into the soil around the roots".[09]

Knowing that the objective of his work might be misinterpreted and that he might be seen as a failure, Harry made the point time and again that his aim was to understand the cause of Coffee Leaf Disease, so that it might be seen how it was possible to prevent future outbreaks of the disease; he was not aiming to find a cure for established infections. He had seen for the first time how a plant disease is spread by airborne spores. He was therefore able to argue, quite correctly, that the removal of severely affected plants together with the pruning of infected branches from less severely affected ones, followed by the destruction of all infected material by burning, would be beneficial, as would manuring in order to promote the healthy fresh growth of the bushes. He saw that planting trees amongst the lines of coffee would form a barrier to check the spread of spores. Almost certainly, the origins of *H. vastatrix* were in the Ethiopian highlands, from where its host originated and where it had over thousands or millions of years reached some evolutionary balance with its host. Possibly it travelled with the coffee plants wherever they were introduced. Thus, the rust had probably been lurking somewhere in Ceylon, or in Southern India, for many years before it was detected in 1869. But it seems that, encouraged by the sudden proliferation of its host, a new, much more virulent, strain had recently been able to evolve. In their desire to reap the greatest profit the planters had failed to leave natural belts of forest between plantations, belts which could have acted as barriers, checking the spread of fungal spores.

Harry was ahead of his time in another way, that is in his recognition of the importance that a lack of host diversity had played in the origins of the coffee rust epidemic. The maintenance of genetic diversity on a local scale is recognised today as being a key factor in the avoidance of epidemics. The science of genetics was still at an embryonic stage in the 1880s – Gregor Mendel's studies of inheritance in the garden pea had been published in 1866 but the work of this obscure Moravian priest was to go unrecognised for another 35 years – so a genetical explanation of the phenomenon was not available to Harry. He saw clearly however that the best way to avoid or check the spread of crop pathogens in future was not to destroy infected material or to plant windbreaks of other species but to avoid large-scale crop monocultures. Modern farmers have learned the lesson; while they will often still grow the same crop species as their neighbours they will, if they are wise, not grow the same cultivar.

In the central highlands of Ceylon in 1880 there was just a monoculture of coffee; there were no other crops and little natural vegetation left to provide barriers to the spread of rust spores. To make matters worse, because coffee bushes carry leaves throughout the year, there was never a time in the year when the fungus was starved of resources: populations of rust fungi infecting annual crops are naturally checked each year during the non-growing season because, as we have seen (p. 9), the fungi are obligately biotrophic and unable to multiply away from the living host tissue. In addition, spread of coffee rust is favoured by warm moist winds. The tropical climate in Ceylon with one 'wet' and one 'dry' monsoon every year meant that conditions were frequently ideal for the transmission of viable spores between areas and across years.

The time that Harry spent in Ceylon was one of rapid personal development. We can begin to learn about his character and qualities by following his experiences through these critical years. Above all, he had to stand alone in both his professional and his personal life; letters could be exchanged with his family or his mentor, Thiselton-Dyer, but it took at least three months for any sentiment, enquiry or question to be answered. Thus, while Dyer remembered that letters from Ward reporting his progress often betrayed moods of depression or discouragement, periods when he needed injections of support and encouragement from the older man, the lack of rapid communication meant that Harry had to come through low periods on his own. With the resilience of youth, he survived.

He formed a close friendship with Henry Trimen, his neighbor in Peradeniya, but there could never be the same intimacy with Henry that he enjoyed with his parents or with his sisters, Elizabeth, Jessie and little Nellie. Letters to his family give nevertheless a clear impression of a young man who was naturally cautious but whose energy, determination and strong character enabled him to overcome any inherent weaknesses. From them, we glimpse a man with a distinct sense of fun, an eye for the absurd, and a talent for music that was often to prove his key to social acceptance. As his time in Ceylon progresses, his letters show a growing self-confidence and greater awareness of the requirements for future success in his career.

Before his appointment as 'The Government Cryptogamist' (an expert on non-flowering plants) – a title from which he took equal pride and amusement – his experience of the world had been limited. Indeed, he had travelled no further than France and Germany. For most of the time on board 'L'Ava' (Fig. 1:4), he had travelled with another young botanist, Isaac Bayley Balfour, who was travelling to Aden and thence Socotra, a small island just off the Horn of Africa, where he was going to do some fieldwork. Balfour was more adventurous than Harry. This might have been because he had more money to spend but it probably had more to do with a greater self-confidence. Isaac was well connected in the botanical world, for his father was a professor of botany and, at the time of the voyage, Isaac had just accepted the Chair of Botany at Glasgow. Many years later, when Harry

Fig. 1:4. MMS 'L'Ava' on which Harry travelled from Marseille to Colombo. The ship sailed the Indo-China route for the french line Campagnie des Messageries Maritimes. Displacing 4420 tonnes and carrying 300 passengers, she was a substantial vessel. The photograph was taken in Port Said some time before 1883. (Reproduced by kind permission of Philippe Ramona.)

had some significant achievements behind him and should have been less self-doubting, he still lacked confidence in Balfour's friendship despite their shared experiences.

On reaching Naples, Harry could hardly wait to tell his sisters of the beauty of the bay and of Mount Vesuvius beyond. He wrote, "As soon as we anchored swarms of Neapolitans clambered from their boats on to the vessel, and one group sang and played some well known operatic airs very fairly, other commenced to turn our decks into a bazaar for cameos, sticks, brilliant toys, etc, etc, some very pretty, but all ruinously dear: Balfour has gone ashore, but (besides the short time being useless unless, as he, one wants a few things only to be got here – the last European point) the Neapolitans will refuse to put you on board again unless you pay them exhorbitant fares, and will laugh while you miss your boat hence I prefer to keep my cash and place – one has not too much of either."[10]

Surviving terrible sea-sickness, Harry arrived at Port Said on 18 January. Horrified by the squalor of town and the untrustworthiness of its inhabitants, he nevertheless ventured ashore together with Balfour.

> "Having quarrelled with the Arab who rowed us ashore as to whether we give him English or French money, who has and who has not paid, etc etc we land: Balfour and I went together with several other passengers. A swarm of filthy young arabs crowd round us, one spoke good English and offered to guide us – we wanted no guide, nevertheless these rascals (i.e. the 2 or 3 who had made

up minds to stick to particular us) never left us, tho' we gave them not a screw and they failed to rob us. We visited the Post Offices, the great Café, and looked at shops etc in the French quartier, then made our way to the Arab town. ... Now picture swarms of the ugliest and dirtiest children playing, quarrelling and yelling with as many dogs, which were all and everyone scratching themselves for vermin – further picture that from each filthy hovel the offal is simply thrown into the 'streets' (courtesy alone justifying the word) that apparently no water-closets exist, for one has to pick every step or be entrapped into a mass of excrementa etc etc, picture all as bad as you can even tho' you had seen Seven-dials and the dirtiest German towns at their worst, and you would still probably fall short of the mark. The foetid sour smell made me feel sick and giddy, and I smoked desperately to guard myself. ... The indiscriminate demands for 'Backsheesh' (alms) on all sides sickens me, and the truest generosity is perhaps to refuse ... Having made ourselves nearly ill, we went back to the ship. ... We had to keep our ports all fastened down and the cabin doors locked on account of their notoriety for stealing: in port, you know, the inhabitants swarm aboard to make sales on deck, and woe betide any valuable they can lay their hands on. Hurray, consider all the world a thief, till you prove him honest.

After dinner we went ashore again and visited the Café, drank coffee and heard 'music' varied by singing and gambling: I did not stake anything, and was much interested in the gambling saloon – in five minutes a passenger lost 115 francs (nearly £50) playing 2 franc stakes. The waiters cheat you by giving you foreign change for English or French money, and it appears a deal of gain is made by melting coin here. ... I was rather tired, and persuaded Balfour to return aboard with me at 10pm: the rest stayed for a bal masque to commence at 12, but since we had no firearms we decided not to run the risk of being robbed, especially as it is not easy to deal with Foreign Officials in Egypt. Once on board I felt more comfortable."[11]

Balfour could be forgiven for thinking Harry was a 'wet blanket'.

Long sea voyages enable the traveller to acclimatise gradually to different climates and, at least on deck, there are usually cooling sea breezes. Having complained of the heat in his cabin, Harry must then have been shocked by the stifling heat as the 'Ava' sailed into the shelter of the long breakwater being constructed around the great harbour at Colombo. The air would have been heavy and, as today, would have carried the characteristic scents of coconuts and cumin. The first impression was one of redness, of red buildings and red streets, all in the quivering glare of the sun. The city and its surroundings were unrelentingly flat, and up and down the narrow streets, in shade or glare, there ran a furious tide of life. Harry was greeted by the local officials who were welcoming and, even if they were not convinced that such a young Englishman sent out from Kew could solve their problems, they were hospitable and helpful. There was no time to lose, however, and much to Harry's relief he was soon able to escape the humid heat of Colombo, taking the four and a half hour train journey up-country to Kandy where the air was fresher and a bungalow awaited him at the Royal Botanic Gardens, Peradeniya (Fig. 1:5).

The gardens were rather less manicured than they are today. Harry saw them as a magnificent wilderness, populated by wild animals, and full of

Fig. 1:5. The Royal Botanic Gardens, Peradeniya. (From The Gardeners' Chronicle, 4, August 1900.)

exotic flowers set among plantings of tall tropical trees whose branches were laden with fruits and fruit bats. A scent of orchids pervaded the air. Professionally, he was introduced to whole families of tropical plants he had never seen before. Privately, he thought he had arrived in Eden. He wrote home that the Garden was as close to Paradise as one could get. In early letters he delighted in telling his little sisters about the huge fruits he was eating (some as big as 'Rip', the Ward family's pet dog), and the huge and extraordinary insects, spiders and snakes he found inside and outside his bungalow.

The first task Harry set himself was to meet the planters, to get their first-hand impressions of the disease and to assess the scale of their problems. On 15th March "a tour through the chief districts in the southern and eastern parts of the Central Province was commenced". He had learned to ride before leaving England and took immense pleasure from his early expeditions into the jungle. He wrote, "'Bob' is my horse … He is a splendid old fellow and I bought him cheap from a man going to England: none of your staid old milkmaids, but a good strong galloping horse".[12]

It is not known whether he travelled alone but he does record "I carried microscope and drawing apparatus everywhere, and, as always, carefully figured any objects of importance to the enquiry". He was certainly entertained by the planters he visited; "I met a piano or two in the jungle, and drank some excellent champagne now and then, you [mother] need not imagine that my steed and I are homeless and friendless in the tropics". Not every planter could afford champagne for, as the extract[01] at the start of this chapter shows, many planters were in a desperate situation. At best the planter's life was an isolated one. Now, some who had come to Ceylon in advance of their families could not afford the fares to bring them out; others could not afford to repatriate their families from the collapsing plantation life. Often, when visiting the poorer planters, Harry had to sleep on couches or boards, sometimes in his clothes for want of a blanket. But, unlike the planters, he could escape to his bungalow situated among coconut and mango trees in a corner of the Gardens at Peradeniya!

The bungalow was, in fact, a two-storey dwelling. On the upper floor, there were two bedrooms, a dressing room, and a room he converted into a laboratory. Downstairs, there were dining room, kitchen and stables, together with a large drawing room and verandah, where he hung his hammock. When not travelling, he settled into an easy routine.

> "At 6 a.m. my boy [servant] brings me tea & toast which I take in bed. Before ½ past 6 I am at work in the cool of the day. At 10.30 I have my bath and at 11 we breakfast. After this smoking, reading and lounging till 1 o'clock when it is too hot to go out, and so I do some sedentary work till 3 when "tiffin" is announced (i.e. tea & toast) then, about 4 p.m., it begins to get cool again and I usually try to get my walk in the Gardens, but generally fail from pressure of microscopic work. At 7 we dine – When I say we I mean Trimen and I: one week we dine and breakfast at my place, the next at his, and so on, except, of course, when I am travelling when my house is locked up – After dinner we drink coffee & smoke till 8.30 or so, and by 9 or 10 I am in bed. [Henry Trimen (Fig. 1:6) was himself relatively new to the Botanic Gardens having been appointed Director in 1879]. We have great swells come here now and again; but I am very glad that Trimen entertains and talks with them. I like to be left with my books and manuscripts, and it is a rare pleasure to me to be so delightfully alone."[13]

Harry and Henry obviously enjoyed each other's company and they were to become lifelong friends. But Harry had a streak of intellectual arrogance in his character and it surfaced in the early days of his acquaintance with Trimen. He wrote home, "His intellectual power I consider below par for my ideal: his persistence in any course determined upon is wonderful. The necessities of his position make the never-failing practice of politic tact sometimes look like virtue: in the abstract I consider policy and tact but words for lying." Trimen may have been "somewhat retiring … with an old-fashioned gravity"[14] but he was far from being an intellectual light-weight. He had edited the *Journal of Botany* from 1870–79 and he was elected Fellow of the Royal Society in 1888. He authored one of the most detailed taxonomic treatments of plants written in the nineteenth century, the massive five-volume *Flora of Ceylon*, which was published between 1893 and 1900 (the work was completed by Joseph Hooker after Trimen's death in 1896). Trimen's appointment to Peradeniya was another that can be

Fig. 1:6. Henry Trimen. As schoolboys, both he and his brother Roland wanted to be natural historians. They drew lots: Henry became a botanist, Roland became an entomologist; they both became Fellows of the Royal Society. (By permission, the Royal Botanic Gardens, Kew.)

directly attributed to Thiselton-Dyer's influence and manipulation; in this case, however, Trimen approached Dyer, who was an old friend from King's College School, asking if there were any congenial colonial scientific posts which would allow him to escape the tedium of his life as an assistant in the botanical department of the British Museum. Dyer knew of just the job!

The world of Imperial power and privilege which Harry Marshall Ward and Henry Trimen shared may be glimpsed in Harry's account to his young sisters (June 1880) of Queen Victoria's Birthday celebrations.

> "On the 28th of May, therefore, we great men had to go in full dress at mid-day and attend the Court Levee at the 'pavilion' or Residency: each gentleman sends in his card and passes thro' a large room at one end of which

is the Governor in full court dress, with all his orders and insignia on. Suppose a few have gone, and your brother's name is called out – 'Mr Marshall Ward' cries the aide-de-camp – Mr. M. W. struts forward, hat in hand (a chimney-pot too, oh girls!), and passing in front of his Excellency Sir James Longden, makes a proper and submissive bow in token of his readiness to do anything for the Queen, from kissing the Governor's boots to – what you choose.

Well, there were some funny people there too. All the native princes pass thro' a ceremony of like import. These 'kings in their own country' are very swells, I can tell ye, and I could not describe the jewels they were loaded with; but I'll give ye a type if I can. He was a handsome, fat, heavy man: travelling upwards, you start with his big, splay, bare feet and ankles. Over the ankles come a wide stiff frill of muslin, continued into baggy drawers tied tight at the frills: over this, fastened all right at the belt and over the chest comes the queerest motley of stuffs you can depict – a rich, fancy light affair perfectly covered with large solid gold coins, & hanging as said. But instead of allowing it to remain as a decent gown tied at the waist, & reaching to the ankles, he has all the front part tucked up again into the waist belt – hence the elaborate view of his muslin drawered legs which one gets in front but not behind – To be plain, he looks very like you would picture one of the little girls with long frilled drawers & short waists we find in the old story-books, if she was holding her skirt up in front in a very un-ladylike manner. But the jewels! Besides the gold & silver all over his clothing, he has rings on every finger, & of these, several had the seals as big as the <u>palm of his own hand</u>! I am stating truth. In his ears are huge gold rings as big as curtain rings, & even each nostril has its jewel. His hat is like that of a "Beef-eater" at the Tower [of London] in shape, and bedizened ad-lib – picture about a hundred such fellows, all in different colours, but sticking to the frilled drawers with obstinate pertinacity.

Very well – we went home & had tiffin & finally dinner. At 9 p.m. the carriage comes, and Trimen and I go off again to court in full plumage for 'the Ball' – the greatest event of the year in a crown colony is 'The Queen's Birthday Ball'.

I suppose there were about 200 or so there – civilians, soldiers, planters, &c. Ladies are scarce always out in the Colonies, but we raised a very fair crowd, & some good dancers. It was luckily a very cool evening or I for one could not have danced 8 dances – However I did, gallops, waltzes, and a polka.

I cannot admire the ways of the Colonies – though I am personally very fortunate, being treated with great respect, and even sought after by the people here – for money is everything with most people. The ladies are very jolly, if you do not mind certain little evidences of Colonial life. All, or nearly all, are married – no pretty girl remains single for long out here. One thing in the women strikes me forcibly: while their beauty, figures, and even to some extent their grace disappears under our torrid sun, they are very honest and well educated as a rule – if they flirt, it is done to some tune you can understand. I suppose things are not quite as satisfactory in India and some other Colonies. I danced with married ladies each time, and have no cause to expect challenges from their husbands: We got home by 3.30 a.m. next day, and I slept 'a few' – my gloves cost me 7/6d and my half share of the carriage 15/–, so you may rest assured I shall not want more than one Q. B. Ball in the year."[15]

It was as well for Harry that his feeling of being intellectually superior to Henry Trimen either disappeared on better acquaintance or was not allowed to cloud their relationship because, ultimately, the way in which Harry's work was presented in Ceylon depended on Trimen. Formal Reports on the Coffee Leaf Disease were presented to the Colonial Government via Trimen, whose duty it was to write an introductory commentary. There was pressure on Harry from the planters, and to some extent thereby from the government, to find a cure for the disease. Harry needed Trimen to understand there were no instant solutions – no quick fix – and to support his programme of research. Trimen did exactly that. In his introduction to the Preliminary Report[16], dated 15 June 1880, Trimen wrote, "... I have noted the solid method of his procedure, and the sure basis he has laid for further progress. I would especially call attention to the *character* of his work, which is that of pure investigation followed out with a single eye to the elucidation of the difficult problem placed before him." In his preface to the Third and final Report, Trimen wrote, "Mr Ward has done well to insist strongly that to find an agent that will kill *Hemileia* is not the most important ... thing to do." He went on to emphasise that practically available chemical substances were not the direction in which relief would be found. If there was a solution, it lay in careful cultivation, judicious manuring and attention to individual trees. The Government accepted the Report although, as recalled by one of Trimen's visiting 'swells', the distinguished German zoologist Ernst Haeckel, this didn't stop Harry being sharply attacked in the local newspapers by some of the coffee growers.

It was Trimen, with Haeckel, who waved-off Harry from Galle harbour as he set sail for home early in 1882. And, when his ship reached Suez, after he had overcome his usual early bout of sea sickness, "I know I have [an immortal soul] because I brought it into my mouth once or twice recently," it was to Trimen, "you dear old fellow," that Harry wrote asking for the latest news from Ceylon, adding "I was very sad at parting with you & Haeckel, and felt very like running after you".[17]

The whole experience, of places and people, and not least the public criticisms he suffered, served to temper Harry's character. His standards remained the highest and his first demand was always that he himself met those standards but, as he remarked to his sisters shortly before his return to England, "No doubt you will remark great changes in me when I return – The one great lesson my travelling has taught me is to take men as I find them, and not feel annoyed if they fall below or pass beyond my standard of excellence; which latter I also discover is always changing in details".[18]

There is an interesting footnote to Harry's time in Ceylon which demonstrates not just his ambition and appetite for work but, overriding these, his fascination with, and love of, fungi and plants. He continued to botanise when off-duty. He was particularly fascinated by the many fungi growing on leaf surfaces in the tropical vegetation. Having ruled them out as the direct cause of the Leaf Disease, he nevertheless wanted to know how they fitted into the microbial community. Remarkably, it was not his writings about coffee rust disease that attracted the immediate admiration of the scientific community but the three papers he wrote describing obscure tropical fungi. He described separately two fungi, *Meliola* and *Asterina*. These belong to a group known today as 'sooty moulds' because their pigmented hyphae build a black weft of threads as they spread over leaf surfaces, feeding on soluble nutrients that leak from the leaf and more particularly on the sugary excretions of the many insects that suck sap from leaves and stems. In a third paper, judged by Thistleton-Dyer[19] as "the crown of Ward's Ceylon work", he reported that another organism of epiphyllous habit, *Mycoidea parasitica*, was a lichen, a mixture of both fungus and alga. Dyer's curious judgement is better understood if we remember that this was the age of Empire, when collecting, recording and describing were of paramount importance in biology. However, even in this work, dull to the modern eye, Harry asked questions ahead of his time. He had the vision to ask whether these harmless fungi might influence other, pathogenic, organisms and, thus, indirectly affect the health of the leaf: the answer is, they do. Today, interaction between organisms on leaf surfaces is a major field of research.

The focus of coffee growing soon moved from Ceylon to Central and South America where, a century later, the rust fungus followed, probably carried accidentally in an airliner. As the crop in Ceylon declined in the late 1880s and early 1890s, plantation workers suffered extreme hardship. In a friendly gesture, the Royal Navy even helped transport some of the Tamil workers back to their homeland. However, some planters avoided bankruptcy and in time replanted the highlands with tea, making Ceylon famous a second time for a major world commodity. The good times returned, at least until recently when, following independence from Britain (1948), the relatively small tea plantations were bought by their competitors, mainly the giant companies of northern India. These have starved the Sri Lankan industry of investment in such basics as fertilizers or new plantings (tea bushes, which are clonal, have a useful life of only 30–40 years and a proportion should be replaced, in rotation, each year). But that's another story!

Today, coffee rust is combated with a combination of genetically resistant cultivars of *C. arabica*, the cultivation of the more resistant species *Coffea robusta*, and a battery of protectant chemicals.

Harry may not have found a cure for the Leaf Disease but the quality of his work, its detailed planning and methodical, painstaking execution, established his reputation among his peers. From Ceylon he had begun a regular correspondence with De Bary. On his journey home, and much to the dismay of his family who could hardly wait to greet him again in England, Harry disembarked at Venice and made the remainder of the journey overland and via Strassburg, where he stayed for about two months. He was able to show De Bary drafts of the papers on coffee rust and *Meliola* that he was preparing for the *Journal of the Linnean Society* and the *Quarterly Journal of the Microscopical Society* and to use the great man as an informal referee. One remark, in a letter written to Dyer from Strassburg on 27 June 1882 is particularly revealing, "I am learning much here, and feel somewhat less inclined to throw up science than I have done lately".[20] Self-doubts had surfaced even at the point when he had achieved so much. De Bary's approval and encouragement to keep going were needed, just as Dyer's had been during the darker times in Ceylon. Harry might have left science at about this time if it had not been for the visit to Strassburg and Dyer's intervention (see chapter 4), and that would have been a tragedy.

Posterity has always associated Harry Marshall Ward's name with Coffee Rust, possibly because of the historical importance of the disease, or possibly because Harry's work put botanical epidemiology on a new footing. Without diminishing this latter achievement, this is to do him an injustice for, as we will see, the latter achievements of his short life were just as remarkable and had greater ramifications. He deserves to be remembered for much more than two years in Ceylon.

Chapter two: Becoming a botanist

"I do not consider it a sine quâ [sic] *non that a professional man must be poor, but I think he & his high thoughts without gold are better than people with narrow, chilly souls & untold wealth. I shall be very glad to get back & hear you play – I sometimes think I should have been happier as a musician, & yet I should have had to be very happy to be so, for I am more than content to be of some real use to Science, & to stand some chance of proving to the world that my mother's eldest son is a man of some intellectual worth".*

Harry Marshall Ward writing from Peradeniya, Ceylon, 15 January 1881, to his mother in Nottingham, England.[01]

The early years of Harry Marshall Ward's life are not well documented. Unlike his son, Frank Kingdon-Ward, he never gained celebrity during his lifetime so those who wrote his obituaries were fellow scientists who, quite naturally, concentrated on his works rather than his life. Few family letters remain, excepting those he sent from Ceylon; others which passed into the possession of his daughter, Winifred, were lost during World War II. What information remains about his childhood and teenage years is therefore somewhat patchy. We may not be able to trace where his love of botany came from, if indeed such an obsession could ever be rationalised, but we can get a feeling for the supportive family atmosphere in which he grew up, and we can see how his brilliance would have been lost if it had not been for his encounters with three men, William Thiselton-Dyer, Sydney Vines and Louis Lucas. His good luck did not end there. As he reached his teenage years, the education system in Britain was at last changing to recognise the importance of science to national prosperity. For at least the exceptionally talented of Harry's generation, there were opportunities for careers in the teaching of science that had never before existed for boys from ordinary families.

Harry was born on 21 March 1854 in the small cathedral city of Hereford set in a rural area still famous for its beef cattle and cider apples. His parents, Francis and Mary Hannah (nee East), were from Lincoln, another small cathedral city some 150 miles to the east (Fig. 2:1). The couple had married in 1853 and Harry was their first child. Francis was one of six men appointed at Easter 1851, from among 40 applicants, as a 'Lay Clark', or 'Lay Vicar' at Hereford Cathedral. He was "to sustain the choral services of the cathedral", the choir having been falling into a state of disgraceful inefficiency. Each clerk was expected to take part in services every day of the year including,

Fig. 2:1. English towns and cities that featured in the life of Harry Marshall Ward.

of course, religious festivals; they were allowed seven days holiday and paid fifty pounds per year (Fig. 2:2). To supplement this meagre income, Francis probably also taught music, for then, and for the rest of his life, he entitled himself 'Professor of Music', a title often borrowed in those times by people who taught private pupils; it did not indicate any academic distinction.

By an odd twist of fate, the Dean of the Hereford Cathedral from 1850 to 1867, and the man who appointed Francis, was Richard Dawes the celebrated educational reformer (Fig. 2:3). A former Fellow, mathematical tutor, and bursar of Downing College, Cambridge, Dawes had in middle age permanently severed his links with Cambridge to marry and become Rector of the parish of King's Somborne in Hampshire. In this agricultural parish of 1,100 inhabitants, he had established a school that was to be a beacon for scientific education in the second half of the nineteenth century. He had no delusions about the abilities of the children in his charge, almost exclusively the sons and daughters of unskilled labourers, but he believed passionately that they, like more privileged children, could benefit from a sound grasp of language. Reading the Bible was essential but he also introduced the famous 'Irish Readers'. These books were astonishingly cheap by contemporary standards and, most importantly, contained both geographic and scientific information. The first four books contained descriptive natural history, rather than what we would regard as true science. By Book 5, however, the largest part is taken up by sections headed 'Physical Geography and Geology', 'Physiology Vegetable and Animal' and 'Natural Philosophy', which included astronomy and chemistry.[02] Dawes seized every opportunity to use

Fig. 2:2. The Three Choirs festival at Hereford Cathedral, 18 September, 1852, in which Francis Marshall Ward would have participated. (By permission of The Dean and Chapter of Hereford and the Hereford Mappa Mundi Trust.)

whatever state aid was available for education. The reputation of King's Somborne spread, together with that of Hitcham school which was developing along similar lines under the guidance of John Stevens Henslow who, like his friend Dawes, had given up a Cambridge post, in this case the chair in Botany (see chapter 7). When the government established the Department of Science and Art in 1853 to reform education in a way that would benefit industry, the two schools were used as models for the whole country.

Dawes also used his school to extend education to local gentlemen and farmers, for example, in 1847 inviting the young Edward Frankland, 32 years his junior, who was teaching at Queenwood College, near King's Somborne, to give six lectures on 'Popular Chemistry (with an agricultural theme)'.[03]

Fig. 2:3. Richard Dawes, Dean of Hereford Cathedral and educational reformer. (By permission of The Dean and Chapter of Hereford and the Hereford Mappa Mundi Trust.)

The two men immediately struck a friendship that was soon to turn into a deep affection. Dawes and his wife, who were childless, virtually adopted Edward, whom Richard always called his son. Edward, his friends and, later, his own children were regular visitors to the Deanery at Hereford, even respectfully attending services at the cathedral. Late in his life and at the pinnacle of his hugely successful career, (Sir) Edward remembered Dawes fondly as "one of the most benevolent and large-minded men that ever graced the Church of England". At Hereford he worked as hard as ever to improve local schools and he taught his flock that greater knowledge of science and the natural world can serve only to magnify the glory of God. It is unlikely that Francis Ward escaped Dawes' influence, or that over the years Francis did not pass Dawes' views on to his growing son.

Francis' duties at the Cathedral did not prevent him from composing and performing music; what little evidence remains shows he had a penchant for the sentimental and romantic ballads that were so popular in the mid-Victorian age. The *Hereford Journal* of 9 May 1855 tells of a concert in which "Mr Ward sang his new song 'The Young Miller' … and accompanied himself on the pianoforte". On 16 April 1856, the same newspaper reported a concert in Abergavenny at which "two pupils of Mr Ward (Mrs Ward and Miss East [his wife and her sister] made their debut before a public audience, and a very flattering one it proved to be". The lay clerks of Hereford Cathedral and several members of the choral body also rendered assistance. A week later, the newspaper reported "Mr FM Ward's concert which came off at the Green Dragon Hotel on Tuesday night last, was not so largely attended as anticipated. It was evident that several of the performers had not fully recovered from the previous night's exertion at Abergavenny, consequently the glees and choruses did not go off so spiritedly as they should have. However, the ladies, Mrs Ward and Miss East, acquitted themselves very creditably, …[reflecting] great credit both on teacher and pupils". At about the time of this last concert, Francis resigned from his post at the Cathedral and the family, now including Elizabeth (Lizzie) born in early in 1885, returned to Lincoln. Harry was two years old.

It is not known why the Wards returned to Lincoln. There is no record that Francis was ever involved at Lincoln Cathedral as a lay clerk or in any other position. On the birth certificate of his second daughter, Jessie, born on 30 April 1857, he again presented himself as a 'Professor of Music', so, perhaps, Lincoln provided more pupils and a better income than Hereford. Harry started school. One report says he was educated at Lincoln Choir School but, sadly, because that school's records were destroyed, this cannot be authenticated. By 1866, when a third daughter Nellie was born on 28th January, the family had moved again, this time to a house in Gill Street, close to the castle and the centre of the city of Nottingham. It appears Harry, now aged ten, attended a private school but again this cannot be authenticated. What happened next certainly can be. When he was 14, the family was faced with financial ruin and Harry was abruptly forced to leave school and to find a job as a clerk. His father, Francis, had unwisely agreed to stand surety for a friend or relation, the guarantee had been called in, and the Ward family suffered as a consequence (Fig. 2:4). It was not unusual in 1870 for young men to leave school at or before the age of 14 – indeed, many never received formal schooling – but for one with professional ambitions, it was likely to be disastrous.

The amount of money that Francis had to pay is not known but the family did move house just before and just after the crisis. They moved only a few streets each time but the size and facilities of the houses most probably reflected the family's fluctuating financial state. The choice of houses must always have been dictated by the Wards' need to have 'live-in' students, who could help the family's finances by paying for their board and lodgings as well as their musical education. Harry's letters from Ceylon suggest, however, that at least some of these young pupils were treated, and loved, as part of the Ward family. Harry had grown up in the expectation that he would follow a career in music, like his father, a career for which his fine baritone voice fitted him well. Although he had a great love of music, which he

Fig. 2:4. Francis Marshall Ward who taught music and composed hymns and anthems such as "I have set the Lord always before me". Photographed in Nottingham. (By permission, the Royal Botanic Gardens, Kew.)

retained throughout his life, it was about this time in his life that he developed a radically different view of his own future. Possibly he turned away from music because he saw that, even in the best of times, it offered only a meagre income. Probably, through his schooling, until it was cut short, and later through the evening classes he attended after work as a clerk, there was growing within him an even stronger love – a love of science. His family did not stand in his way. Indeed, both his parents seem to have had open, enquiring minds and to have encouraged discussion and learning with their children. His mother had a keen interest in geology and taught Harry about fossils during the family's holidays at the Yorkshire seaside town of Scarborough, where fossil remains can so easily be seen in the cliffs and rocks of the beaches. The fossil record, evolution and Darwinism must have featured in the family's discussions. It was at about this time he joined the Nottingham Naturalists' Society. The Society was founded in 1857 and the earliest surviving records of membership are for 1875, when HM Ward was listed as a member of the Microscopical Section. Through its talks and excursions, and by meeting other members of the Society, he would have become familiar with the common plants and animals of the East Midlands. He may already have been familiar with exotic trees since the family homes were close to the Nottingham Arboretum, a place of cool refuge on hot summer days.

A life in scientific research could have been no more than a dream. In botany, there were, outside Kew and the botanic gardens, only university posts to think of. These were very few and were often filled from within a closed network of men with whom neither Harry nor his family had any connection. There is no record of a member of his family ever having gone to university before him. Pragmatically, he must have known that the most realistic prospect for him, if he wanted a career in science, was not as a researcher but as a teacher of science in schools, one of the small band of men who were to be in the vanguard of educational reforms about to sweep the country.

Harry was born only three years after the Great Exhibition of 1851, an event which was the brainchild of Albert, the Prince Consort to Queen Victoria. Judged an unparalleled success by the public, the Great Exhibition sowed seeds of disquiet in the minds of some discerning observers, including Prince Albert, a number of leading scientists with whom he associated, such as Lyon Playfair, and a small but growing number of industrialists. It was becoming clear to these men that deficiencies in the educational system threatened the continuation of Britain's domination of world trade and commerce. Its lead over Germany, France and the United States was slipping, a message painfully rubbed-in by the second International Exhibition in London in 1862 and the Paris Exhibition of 1867. In 1868 reports by Matthew Arnold (Inspector of Schools and famed poet) and by a House of Commons Select Committee under Bernhard Samuelson (an industrialist), it was argued that the commercial and industrial success of Germany was founded upon the greater availability and higher standards of education in German Schools and Universities. In agriculture too, Germany was leading the world; the great chemist Justus Liebig was showing how scientific principles could be applied to the study of fertilizers and the first tax-supported Agricultural Experimental Station was founded in Saxony in 1852. The latter was, in itself, only of local importance but it was indicative of the recognition in Germany of the need for better educated farm workers as the mass migration of labourers from rural communities into the cities gathered momentum. The very first issue of *Nature*, in 1869, reflects the admiration of Germany's educational system felt by informed British scientists.[04] All agreed, it was time Britain learned from the German model.

The extent of the British problem can be seen by examining opposite ends of the educational spectrum. At one, elementary education was neither compulsory nor universally available: at the other, Oxford and Cambridge, those ancient universities from which politicians and intellectuals were almost exclusively drawn, were steeped in conservatism, and in the traditional disciplines of the classics and mathematics. Indeed, such an influential figure as William Whewell, Master of Trinity College and *himself* a mathematician and mineralogist (not to mention a philosopher of science), could support the introduction of a Natural Sciences Tripos while believing that the teaching of 'progressive' subjects such as geology or chemistry should only supplement the teaching of 'permanent' subjects such as geometry, mathematics and the ancient languages.[05] He considered the former subjects were unsuitable for academic instruction because students might lose reverence for their teachers when they discovered Professors could not maintain doctrines as indisputable as mathematical or arithmetical truths. It was said of Whewell that he felt a century should pass before new discoveries in science were admitted into the course of academical instruction.

In 1870, just as Harry was forced to leave school and to look at his future with dark foreboding, a chink of light appeared. The pressure for educational reform achieved its first tangible successes. The Elementary Education Bill was driven through the House of Commons by WE Forster (Matthew Arnold's brother-in-law) on behalf of the Whig administration of WE Gladstone. Through the establishment of School Boards, which would provide education in areas where it had not previously been available, such as the rapidly growing cities, the Bill provided the opportunity of elementary education for all. Acts of parliament in 1876 and 1880 finally made education compulsory between the ages of 5 and 10. Also, in 1870, the Royal Commission on Scientific Instruction and the Advancement of Science was appointed under the chairmanship of the 7th Duke of Devonshire.[06] 'The Devonshire Committee', as it became known, sat for five years and produced eight reports. Apart from its criticisms of the old universities and encouragement of newer institutions such as Owens College, Manchester (of which, more in chapter 4), the first and second reports included recommendations that together were directly to affect Harry Marshall Ward. Those in the first report related to the future of the Royal School of Mines, while those in the second related to classes taught at the Department of Science and Art at South Kensington, west London.

The relevance of mining was that it had provided Britain with the coal and minerals upon which the industrial revolution had been founded. A Government School of Mines had been opened in Jermyn Street, central London, in 1851, and in 1853 it absorbed the Royal College of Chemistry. Among the subjects taught were mining and mineralogy, metallurgy, geology, chemistry and natural history. Among its staff were, at different times, such distinguished scientists as Lyon Playfair and Thomas Henry Huxley. The School of Mines was soon to fall under the aegis of the Department of Science and Art, itself founded in 1853 with the object of raising industrial standards through education, but its main problem was that it failed to attract students in significant numbers. Up to 1859, it had an average of only 14 full-time and 51 part-time students. Numbers only picked up in 1859 when the Department of Science and Art (otherwise known as 'South Kensington') started a scheme to license schools, or classes of 10 or more students, throughout the country. Teachers of licensed classes had to be of an approved standard and, to induce them to reach this level, their salaries were to be raised in line with the qualifications they obtained. There were annual examinations for students and their teachers received additional payments according to the results; £3 for a first class pass, £2 for a second, etc. Thirty classes and 1,340 candidates were subsidised in 1860, the first year the scheme operated, and by 1873 there were 1,182 classes and 24,674 candidates. The system failed because syllabi were theoretical and, most disappointing to employers, no manual training was included. In retrospect, the timeless educational truth can be seen: a system of payment by results can lead all too easily to stale, unimaginative teaching.[07]

The Department of Science and Art had consistently refused to make supplementary payments to teachers in elementary schools, in effect discouraging teachers from gaining qualifications and excluding most of the children in school from any experience of science. Recognising the potential waste of talent resulting from this policy, the Devonshire Committee recommended that science should be taught in all elementary schools and in colleges, thus immediately increasing greatly the nation's need for qualified science teachers; it also recommended closer inspection and better funding for approved courses. The Committee opened the way for Thomas Henry Huxley and other reformers to take teaching of chemistry, physics and natural history (= biology) away from Jermyn Street. Teaching was moved in 1872 to a new building at South Kensington which had been intended for the School of Naval Architecture. The Science School, which was thereby founded, was later re-named the Normal School of Science (after the French Ecole Normale) (1881), the Royal College of Science (1890), and the Imperial College of Science and Technology (1907). At the outset, its express purpose was to train teachers of science. Students were selected on the basis of their results in examinations set by the Department of Science and Art; those who were successful were admitted for free instruction and they received a maintenance grant of one pound and one shilling per week. The promise of this modest supplement to whatever they might be able to save for themselves was enough to persuade young men like Harry to set their sights on a career in teaching. (The author HG Wells was a student science teacher a few years later and his *Life and Loves of Mr Lewisham* gives a unique glimpse of life at the Normal School.)

Accounts variously describe Harry as attending evening classes in Nottingham, Manchester and South Kensington during the years 1872–73, but the most likely explanation is that he took the classes approved and examined by the Department of Science and Art while continuing to live at home in Nottingham. He could probably not have afforded to leave home earlier. His talent and dedication to study was rewarded; he won a place as a full-time student at the Science School, starting in the 1874–75 session.

Luckily for him, Harry arrived soon after the interior of the buildings at South Kensington had been significantly re-furbished. Huxley's new design included laboratories that enabled him to introduce practical teaching to back up his lectures, a method of teaching he had long been convinced was essential to the proper education of a scientist.[08] The approach was totally new in Britain and just as revolutionary as the theory of evolution which Huxley was busy defending on Charles Darwin's behalf. For six weeks in the summer of 1872 Huxley taught a vacation course for teachers in which he gave the daily introductory lecture at 10.00 a.m. – probably emphasising the unity of zoology and botany, which was central to all his thinking – before students moved into the laboratory for two sessions, 11.00 to 1.00 p.m. and then 2.00 to 4.00 p.m. In spite of the shortage of equipment in that first year, the course was enormously successful and it was agreed that it should be repeated the next year. However, the demands of organising the move to South Kensington, involving the planning of laboratories as well as devising of the new courses, plus the demands of being public spokesperson for the 'evolutionists', had taken its toll on Huxley's health and he was forced to take an extended holiday. To cover his courses, and to strengthen the teaching with a specialist botanist, which he was not although botany had been his first love, Huxley appointed William Thiselton-Dyer to teach the botanical parts of the course.

It was as student and teacher, respectively, at South Kensington that Ward's and Dyer's paths first crossed. Later, at the major turning points in Harry's career, it was always Dyer who was there, in the background, clearly pointing the direction which Harry should follow. Whether those directions were in Harry's best interests will be examined later (chapter 8) but at this point we need to understand how Dyer happened to be in a position to direct events in the life of Harry Marshall Ward – and the lives of many other young botanists of the time. It will be helpful also to get some feeling at this stage for the personality of this strong-willed man, for then we will be better able to understand how, at a personal level, he was able to exert control over others.

Born in 1843, the son of a prosperous Westminster physician (and grandson of a barrister who obtained a Royal Licence to add Dyer to his own name Thiselton), William Thiselton-Dyer was educated at private schools and went on to King's College, London, with a view to a career in medicine. Medicine was soon abandoned however when he switched to Oxford University, where he eventually graduated with a second class degree in Mathematics and a first in Natural Science (1867). When his father died in 1868 he was in need of a job. He had been 'promised' the chair at Oxford but to his disappointment was ruled out because he was 'not of MA standing' [he had not completed the qualifying number of post-graduate years].

Soon, however, he was appointed Professor of Natural History at the Royal Agricultural College, Cirencester, Gloucestershire. While there he collaborated with a colleague and chemist, Arthur H Church, in writing an English adaptation of Samuel W Johnson's American book, *How Crops Grow*, which as its subtitle states was written for 'Agricultural Students'. The book is heavily influenced by Justus Liebig's published works with much emphasis on chemical analyses of the ash of plants because, as Liebig argued, ashes reflect what plants need from the soil. The book nevertheless contains important sections on the structure and life of the plant, the latter including what we would regard today as physiology. The revised version demonstrates an awareness of recent work coming out of German laboratories. There is, for example, a section on 'Causes of the Motion of Juices' including descriptions of Wilhelm Hofmeister's conclusions on the subject and there is, in some detail, a discussion of the apparatus used by Julius von Sachs to demonstrate osmosis and the ascent of sap. Quite remarkably for its time, the book acknowledged a link between plant physiology, widely viewed as an academic discipline and detached from practical matters, and the real day-to-day problems of arable farming. This fresh new approach or attitude must have been reflected in Dyer's teaching and it is the first expression of the educational revolutionary he was to become.

Dyer also found time during this period to publish with his old school friend, Henry Trimen, a *Flora of Middlesex* (1869), a project on which they had embarked many years earlier. Rural Gloucestershire was not exactly to Dyer's liking and in 1870 he was able to escape to the Royal College of Science in Dublin, moving on, in 1872, to the Science School at South Kensington where in 1874 he first encountered Harry Marshall Ward, a man only 11 years his junior.

During his time in Dublin, Dyer had begun an exchange of letters with Joseph Hooker, Director of the RBG, Kew, and on Hooker's recommendation, Dyer was appointed Professor of Botany at the Royal Horticultural Society (RHS) on his return to London. The Chair made minimal demands on his time so, in addition to his duties at South Kensington, Dyer was able to be employed as Hooker's private secretary for "a few hours each day, four days a week".[09] The arrangement worked in his favor and in 1875 he was named Assistant Director of Kew. In 1877 he married Hooker's eldest daughter, Harriet Ann, and in 1885 succeeded his father-in-law as Director (Fig. 2:5). Dyer always claimed he would have preferred to stay at South Kensington if a suitable post had been offered to him, but at Kew he found his metier. He was amongst Kew's most successful Directors. He rationalised its administration and radically re-landscaped its grounds, but his greatest contribution was to develop Kew's role at the centre of the botanical web which, as we saw in chapter 1, extended throughout the British Empire. Throughout the colonies there was scarcely an appointment to a Directorship of a Botanic Garden, or even to a post of Head Gardener, that was not given to an ex-Kew man or, at least, approved by Kew. The botanists of Kew were intimately involved in the exploitation of native plants and the movement around the world of crop plants to be introduced to new settings, with Kew supplying plants and seeds, and giving guidance on everything from cultivation and the occurrence of major pests and diseases, through to marketing.

An authoritarian by nature, Dyer demanded unquestioning loyalty and obedience from his staff. He imposed order on the establishment by making

Fig. 2:5. William Thiselton-Dyer, Director of the Royal Botanic Gardens, Kew. He designed uniforms for the garden staff and for himself. (By permission, the Royal Botanic Gardens, Kew.)

everyone wear a uniform, the design of which reflected the nature of the job they did and their position in the hierarchy. He even designed a Director's uniform for himself, which suited his military bearing and looked as though it was strongly influenced by his earlier links with the Oxfordshire Light Infantry – as a break from teaching at South Kensington, he had liked to go on manoeuvres with the infantry in the wilds of Dartmoor. Described by a contemporary as 'pompous and popish', many of the changes he brought about can be seen in retrospect as liberal and enlightened. Thus, his courses at South Kensington were the first to include female teachers and under his regime at Kew the first women gardeners were appointed, although they had to wear strictly unalluring and unprovocative uniforms. Such changes may have been pragmatic, reflecting his dedication to progress rather than any sympathy for women's emancipation, but their effect was positive.

The young Beatrix Potter fancied Dyer was a misogynist. She was a keen mycologist and lichenologist before she ever started writing the Peter Rabbit books for children and was introduced to Dyer by her uncle, the chemist, Sir Henry Roscoe. She sought help with her studies of the germination of spores of fungi of the Agaricaceae (mushroom family) but was squashed by Dyer whom she found "a thin, elderly gentleman … with a dry, cynical manner, puffing a cigarette, but wide awake and boastful. … His line was on the outside of civil."[10] Thankfully, Beatrix did get help from Kew's resident mycologist, George Edward Massee, who later presented to The Linnean Society a paper describing her work, women being barred from addressing the Society. Massee was himself a Dyer appointment; in somewhat acrimonious circumstances he replaced Mordecai Cubitt Cooke, an older man who typified an earlier age at Kew. Cooke's daughter, perhaps not unbiased, wrote, "He [Dyer] started to organise the Herbarium and the old friendly scientific atmosphere was wiped out. He, no doubt, laid the foundation of the efficient Government Department which it became, more suitable for the changing times."[11] In summary, Dyer must have been rather like a hard-nosed, modern, mover-and-shaker. He knew his staff didn't like him but he didn't care. He was fearless in doing what he believed was best for the institution. He was tactless but brave, on occasions prepared to pick a fight with Members of Parliament and even members of the Royal Family. No wonder Harry Marshall Ward, and many botanists of his generation, followed Dyer's directions!

Now, to return to South Kensington in 1873. Dyer later recalled,

> "[Lawson] and I determined, with an audacity at which I almost shiver, to demonstrate to our students things which for the most part we had only seen ourselves in books, and which were regarded with awe as only within the competence of the greatest masters. … We determined to attempt a course of instruction which would embrace the leading morphological facts of every important type in the vegetable kingdom. We, in fact, resolved to adopt exactly the same plan of work [lectures supplemented by 'hands-on' laboratory work] as Professor Huxley in his own teaching had found convenient for the animal side of morphology. …
> "As far as I am aware, no previous attempt had been made in this country to give an extended course of botanical instruction of this kind. … Lawson and I found our difficulties scarcely less considerable than those of the students."

Before classes, the two men often spent half the night arranging and illustrating the plant and fungal material they had acquired, sometimes from contacts far away in Europe. Thus, they demonstrated for the first time in Britain an active plasmodium of a Myxomycete (Slime Mould), they showed their students the motile male gametes of a green alga (*Chara*), and they and their students cut thin sections of Angiosperms (flowering plants) and Gymnosperms (flowerless higher plants) to be studied under their microscopes – in the 1870s there were no microtomes to cut sections automatically.

Nevertheless, Dyer and Lawson, assisted in 1874 by Sydney Vines, found that the novelty of their methods enthused and stimulated their more skilful students. None was more able than Harry Marshall Ward, who made an immediate impact on his tutors. Dyer remembered, "Vines and I soon discovered that we had got hold of a man of exceptional ability. In the examination at the end of the course he soared away from all the other students and stood alone at the head of the first class." Dyer follows with a curious and, were it not for its ominous portents, amusing remark. "The strain of the work on both teachers and students was certainly severe: there was the keenest competition amongst the best men to respond to the demands made upon them, and the tension was sometime rather acute. On one occasion Marshall Ward fainted at his work from no other cause I think than over-excitement". It seems most likely Harry had not eaten recently and this might even have triggered the first effects of diabetes, the disease which was to kill him thirty years later. However, as Dyer later recalled, a stiff drink was administered and Marshall Ward was soon back at work.[12,13]

We know little of the South Kensington years from Harry's point of view. Undoubtedly the science was immensely exciting. He must have been stimulated, and felt privileged, to be moving in the world of men as famous as Huxley. Life was hard, however, as he confessed to his wife many years later. He lodged at some distance from South Kensington and had to walk the four miles each way each day. Walking back on summer evenings he would look up the stars and revise his astronomy, which was a part of the science course, but in winter the long trek was a cold, often wet, and usually miserable experience. Nevertheless, from time to time he managed to escape to his family and the home comforts of Nottingham, where he even found time to address the Nottingham Naturalists' Society on subjects ranging from 'Early stages in the development of the bird' to 'Reproduction in seaweeds'. Despite being an inexperienced public speaker, he had the arrogance of youth to tackle subjects as profound as 'Fact and theory in natural selection'.[14]

In the autumn of the 1875, Harry went to Manchester to study for one term at Owens College, the forerunner of Manchester University. What caused him to go to Manchester is not certain but reason suggests it was at the invitation of Louis Arthur Lucas, a friend he had made at South Kensington. Lucas' father, Philip, was a leader of Manchester's thriving Jewish community, having built a successful shipping and insurance business before his death in 1865. Perhaps even more remarkably, because of what would normally have been the obstacle of Gentile hostility, he was elected to the City Council in 1851. The Lucas family was among a small number of Jewish 'suburban plutocrats' in Manchester; they had seven servants at their home, 'The Temple', in Cheetham Hill.[15] But they also had strong family connections in London. Louis' uncle (of the same name) lived there and two of his five older sisters settled there after marriage. Young Louis was sent to University College School, possibly lodging with one of his sisters during school terms, and when he was old enough to leave school he undoubtedly spent a lot of time socialising with family and friends in London, enjoying the attractions of what was at the time the world's greatest city – as any young man from the provinces would wish to do. In London he was also able to indulge his interest in science, which seems to have stemmed from an early age for, at 20, Louis had been elected to membership of the Manchester Literary and Philosophical Society. We know that Huxley's fame and his reputation as an innovative teacher attracted to his classes a number of casual students, and among these was Louis Lucas. He was not formally enrolled as a student at South Kensington but rather he had an arrangement whereby he attended classes at his whim and often shared a microscope with Harry. What would be more natural then that Louis should invite his new friend back to Manchester.

At Owens College, Harry would have been able to advance his knowledge on many fronts under the guidance of one of the most celebrated scientists of the day, William Crawford Williamson, who held the chair of Natural History, Botany and Geology. Williamson's academic status was high both within the College and nationally and, through his medical work as well as his many years of service to the College, he had built up a wide range of influential friends, not least Solomon Levi Behrens. This old man was a rich benefactor of Owens College, and one of his grandsons, Edward, was married to Louis Lucas' sister, Abigail. Probably, therefore, Lucas was able to make the right introductions for Harry on his arrival at Owens. Certainly, Harry became a firm friend of Edward Behrens, regularly writing to him from Ceylon to tell him in the greatest detail about the progress of his research.

Harry's meeting with Lucas and his subsequent visit to Manchester were to prove pivotal to his personal as well as his academic life for it was during this first stay in the city that he somehow met Selina Mary Kingdon, his future wife. Perhaps the connection was made via Owens College since Selina's father, Francis, was a Demonstrator in its Department of Natural Philosophy (Physics).[16] He was the only member of staff without a degree and quite why he had taken that post in 1870 is a mystery. Although Francis was related by marriage to the brilliant but eccentric mathematician, William Clifford Kingdon – who once hung by his toes from the weathercock of a church tower and whose ideas on the relationships between energy and matter would not have been out of place in the twentieth century[17] – there is no evidence of Francis' having a scientific career before or after his period at Owens. The family had strong connections with Devon and both Francis and his father before him owned a lace and fringe-making business in the High Street, Exeter, where Selina was born. Perhaps it was the decline of the lace making industry which caused Francis to take his wife and five children from their roots in sleepy Exeter to the huge dirty city of Manchester. What is on record is that Francis resigned his Demonstratorship at Owens in 1876 and the next we know of him is that he died in Texas in 1881, much to Selina's dismay as we will see in chapter 5. It is not known exactly when Francis Kingdon left for America but it was probably after he had given his blessing to Harry and Selina's engagement for it was believed by their daughter that her parents were engaged for seven years before their marriage in 1883.

Not diverted from his studies by romance, Harry carried off prizes in every subject he took up at Owens. He was first in his class in junior mathematics and also in chemistry. He supplemented the daytime learning with evening courses, coming first in botany and second in physiology. Harry was to renew his connection with Owens and with Williamson later in his life (see chapter 4) but by 1876 he was back at South Kensington, this time as a demonstrator, a paid laboratory teaching assistant.

Much as he loved teaching, there must have been a cold fear growing inside him because what he wanted most of all was to be a researcher in his own right, and to be that he needed a University education – something neither he nor his family could afford. What happened next was that his friend Lucas became his benefactor and guaranteed to support him through university. Exactly how it happened is shrouded in mystery and suffused with tragedy. Lucas had long dreamed of an expedition to Africa. Finally, after many delays, in the first months of 1876 he set off across the Sudan accompanied by nearly 200 porters and over 400 cases. Even before he set off it was apparent to all those with any experience of Africa that his constitution would not stand the journey. As predicted, he hardly got beyond Khartoum before his health broke down. General Gordon, whose word was law in those parts, peremptorily ordered him to give up the expedition in fairness to those who travelled with him. But it was too late and, after great suffering, Lucas died on about 16 November 1876 aboard the S.S. 'Massowah' making the short crossing of the Red Sea from the Sudanese port of Suakin to Jeddah in Saudi Arabia.[18] He was only 25 years old.

Harry never acknowledged there was any special friendship between himself and Lucas, maintaining he did not know the identity of his benefactor until after Lucas' will was published in December 1876. When asking advice from Thiselton–Dyer and Vines he said merely that an anonymous benefactor had lodged sufficient money with Mortlock's Bank, Cambridge, if he would take a place at the University. It is possible he thought the benefactor was Edward Behrens, who had clearly taken an interest in his career. However, given Harry's lack of rich friends or relations, it seems Behrens would have been the only alternative, and, given his actual friendship with Lucas, he must have had a pretty good idea it was the latter who was his patron. It remains a mystery why he was unwilling or embarrassed to acknowledge publicly Lucas' friendship and support. Possibly he was afraid people might think he had used Lucas, cynically cultivating the relationship for his own purposes; possibly he would have been embarrassed to acknowledge his support was from Jewish sources for there were still strong public prejudices against people of that race. Whatever, following the encouragement of Dyer and Vines, Harry successfully sat the scholarship exam at Christ's College and, with his tuition fees now guaranteed, he 'went up' to Cambridge in October 1876.

For one so young, Lucas had been an excellent judge of character. Not only had he spotted Harry's latent genius but he had also spotted his man's weakness; if Harry failed to control his workload, any achievements and success could all too easily be undermined. Harry's enthusiasm and dedication were a danger, to himself. Writing in November 1875, when Harry was furthering his education at Owens College, Lucas urged him "Slow and sure wins the day. Work ten or even eleven hours a day. You can do that, if you keep regular hours, for indefinite time. Always take at least seven hours sleep. For the more the work the more the rest required". Tragically, Harry was to forget the last part of his friend's advice!

The Natural Sciences Tripos at Cambridge was broadly based with the Sciences being examined separately in the 'Advanced' part of the Tripos. Teaching was open to all students, offering an ideal opportunity for a man with interests as wide as Harry's. Thus, he attended lectures in human physiology given by the charismatic teacher Michael Foster, an experience for which he remained grateful throughout his life and which strongly influenced his outlook on plants.[12] He also took a special interest in entomology, an interest which served him well in his later career as a plant pathologist and a subject on which he frequently corresponded with his academic friends. Most significantly, while at Cambridge, he was introduced to the world of German plant science. Possibly he had heard a little of it before through Sydney Vines, the same man who had taught him in South Kensington. But now Vines was his tutor in Christ's College and he was able to give more time to Harry and the other aspiring young botanists. Significantly, he was able to bring them the latest news from the German laboratories. Vines was a well-travelled young man who had spent part of his childhood at school in Germany so his knowledge of the language was exceptional. Probably because of this, he more than other young botanists was keenly aware of the 'New Botany' that had been launched in Germany. During his South Kensington years he had translated into English Sachs' *Textbook of Plant Physiology* (1868: translation 1875). But this was not enough for Vines, he knew he had to find out more and the best way to do that was at first hand. So, newly appointed Fellow and Lecturer in Christ's, he had spent much of the first year of his Fellowship studying in Germany. Not surprisingly, he went to Sachs' laboratory in Würtzburg.

Returning to Cambridge for the autumn term of 1877, Vines instituted the first practical classes in botany in the University. In a room lent to him by Michael Foster and with microscopes and other apparatus he had paid for out of his own pocket, he brought together students of botany from the different years. His enthusiasm was not only unquenchable, it was infectious. His small band of followers was given confidence in their own powers of observation, they were taught to challenge the orthodoxy of the time. It is a tribute to Vines' inspirational qualities that Harry and his classmates, including Frederick Orpen Bower, became such influential teachers in their turn. In his own old age, Bower wrote about his undergraduate days in Cambridge, "one still remembers the dilating of the nostril and the missionary aspect of Ward as he spoke of 'The Cause'".[20] Sadly, Vines was to lose his fire in his senior years and, as Professor of Botany in Oxford, was best remembered for being something of a recluse, interested mainly in his own garden on Headington Hill. Harry never lost his fire and, as later chapters will reveal, he and other disciples of 'The Cause' succeeded eventually in their aim, that of establishing in Britain a school of experimental science equal to anything seen in Germany. Their triumph was that as the plant sciences evolved in Britain they took from Germany its leadership of the botanical world.

To return to the late 1870s, although hampered by the university's requirement that he should learn Greek, of which he had no previous knowledge, Harry's academic brilliance and dedication led him, not surprisingly, to a first class degree in the Natural Sciences Tripos of 1879. HM Ward was the only botany graduate of his year.

On 20 November of the same year a paper by H Marshall Ward, BA, Scholar of Christ's College, Cambridge was read to a meeting of the Linnean Society, and subsequently published in its *Proceedings* in 1880. Entitled, 'A contribution to our knowledge of the embryo-sac in Angiosperms', it was the first piece of original research he published and no doubt – as first publications always are – a source of great pride. The work on which the paper was based had been done during the summer of 1879, mainly at the Jodrell laboratory in Kew. Again, Harry had benefited from work of the Devonshire Committee for, in its fourth report, it had recommended there should be a facility at Kew for the study of 'Physiological Botany'. The recommendation would have been lost for lack of enthusiasm if it had not been for the energetic support of Thiselton-Dyer and a generous donation from TJ Philips Jodrell, a friend of Joseph Hooker, which enabled a small laboratory to be built and equipped: it opened in 1876. The subject of Harry's paper was the pattern of cell and nuclear division that leads to the formation of the mature embryo sac. This was a difficult subject upon which to launch a career for, as it is now appreciated, the pattern varies considerably between species, but Harry's conclusions were more right than wrong. From an historical viewpoint, most notable was the way in which he, a novice, was prepared bluntly to criticise the published work of Vesque, an established and well respected authority on the subject. Such boldness in arguing against the accepted position was to characterise many of Harry's later writings, as was his support of those arguments by the presentation of an overwhelming amount of data (in this case superb drawings of the ovular anatomy of *Butomus umbellatus* and eight other species), and a love of wordiness. In short, Harry's writing was never short.

With a passion for 'The Cause' burning inside him, Harry decided that, like Vines, he must experience the wonders of the German laboratories at first hand. So, after graduation, and no doubt with introductions made by Vines, Harry organised a short visit for himself to Würtzburg. Sachs was a scientist of the very highest distinction whose list of discoveries is quite astonishing to modern eyes. Among his 'firsts', he established (with Knop) the essentiality to plant growth of most of the macronutrients, such as nitrogen, potassium and phosphorous; he showed that the pigment chlorophyll is not distributed evenly throughout green cells but is localised in specialised bodies (later called chloroplasts), and that starch is the first visible product of photosynthesis; he showed that the upward movement of water in vascular tissue occurs through non-living cells (xylem vessels and tracheids); and he established the basis of both chemotropism, for example, the growth of pollen germ tubes is directed from the stigma, through the style, by a chemical gradient towards the ovary, and phototropism, the phenomenon in which stem extension and curvature are affected by light. Although Harry did not like Sachs, like others of his generation finding Sachs "a narrow and intolerant man, who rarely stirred from his laboratory",[21] his experiences of that laboratory and the people he met there were to raise for ever his consciousness of the physiology of plants. To help others to the same awareness, Harry published in 1887 a translation of Sachs' *Lectures on Physiology*. The greatest tribute he paid to Sachs was however in his own work, bringing a physiological perspective to plant pathology and, as we will see later, helping to establish another new discipline – physiological plant pathology.

The state of botanical teaching in British universities in 1876, as Harry was about to start his undergraduate studies, is encapsulated in the entry for botany in the 9th edition of *Encyclopaedia Britannica* written in the same year by Professor Hutton Balfour (Isaac's father). Botany was studied, as under Balfour in Edinburgh, in order to give medical doctors correct knowledge of the plants which were the source of drugs. The eighty-page-long article is devoted almost exclusively to flowering plants, whose structure and morphology it describes in such exhaustive detail that the reader can build a mental picture of the plant, providing that is that he or she is familiar with the meaning of the many, many Latin terms and phrases which litter the article. Vines' teaching, and Harry's first-hand experience of the vibrant, self-assured laboratory in Würtzburg, would have been like oxygen to a suffocating man. They would have shown the young man that his chosen subject, botany, was worth studying for its own sake. He could take pride in becoming a botanist.

Chapter three: German lessons

"By 1853 De Bary, who had climbed the intellectual Mount Olympus, finally convinced the counter productive thinkers of his day that fungi could cause disease, and hence, that disease could be controlled by controlling fungi. De Bary established a new milieu – the vitalists now gained the upper hand."

JG Horsfall and EB Cowling (1977) *The Sociology of Plant Pathology*.[01]

Harry Marshall Ward's undergraduate studies at Cambridge and, in particular, the teaching of Sydney Vines would have made him well aware of the writings and the reputation of Anton De Bary. Mycology and plant diseases were high on the list of Harry's interests so it was not difficult for him to decide that while in Germany, visiting Julius von Sachs' laboratory at Würtzburg, he should take a little of his precious time to go to Strassburg (= Strasbourg), where he could visit the laboratory of De Bary, the other great German giant of the New Botany. Accordingly, arrangements were made and Harry, together with two fellow students from Sachs' laboratory, travelled west to Strassburg late in the summer of 1879. One of his companions on the trip was Francis Darwin, with whom Harry soon struck up a lifelong friendship and with whom many years later he was to work as a colleague in Cambridge. Francis had been in Sachs' laboratory the previous summer learning how Sachs studied plant tropisms (movements to or away from light, gravity, chemicals, etc), the same subject which Francis was investigating with his father, Charles, in the somewhat isolated environment of his father's home, Down House, in Kent. As Harry soon recognised, Sachs treated Francis poorly, typically retaining an air of superiority and giving the Darwins little credit for the outstanding work that they were doing, so it must have been a relief for Francis to escape to the much more welcoming atmosphere of De Bary's laboratory.

This chapter is largely about De Bary and his laboratory in the Strassburg Institute (Fig. 3:1) for two reasons. Firstly, because it was he who was responsible for the state of knowledge in plant pathology that was current at the time Harry was embarking on his researches. The methods that Harry was to employ in Ceylon where, as we have seen in chapter 1, he successfully traced the infection cycle of the coffee rust pathogen, followed closely those laid out in De Bary's classical investigations of potato blight. Secondly, because it was in De Bary that Harry found his personal inspiration: he put De Bary in the pantheon of his gods – placing him alongside Goethe and Mendelsohn – but he found also in De Bary a kind friend and patron, who

Fig. 3:1. The Strassburg Institute where Harry visited De Bary's laboratory. The low, ill-lit buildings were those of the old French Académie. It adjoined a Botanic Garden described by Bower[13] as "little more than a closed yard". (Reproduced from Sparrow, FK, 1978, Mycologia 70, 222–52, by permission of The Mycological Society of America.)

would guide and support his researches. As we will see in later chapters, it was Harry's tragedy that he did not always manage to emulate his idol's ability to inspire younger men, or to keep a healthy balance between his professional and his private life.

The passage of time has done nothing to diminish De Bary's scientific reputation – rather, it has enhanced it. In 1977, the distinguished American phytopathologists, James Horsfall and Ellis Cowling, with a sense of fun and tongues firmly in their cheeks, reviewed[01] the genealogy of Presidents of the American Phytopathological Society, the largest professional body of its type in the world. Recognising that excellence runs in scientific families – that Nobel Prize winners are chosen and taught by Nobel Prize winners – they identified four noble families among North American plant pathologists. The ancestral starting point for each was Anton De Bary (Fig. 3:2). De Bary's personal reputation also stands high, letters and obituaries written by his colleagues and ex-students repeatedly stressing his human qualities. They provide moreover a wonderful insight into life in and around a late nineteenth century laboratory; they remind us forcefully that while the concepts and tools of biology may have changed almost beyond recognition in the last hundred years, the amusements and human interactions of scientists have not.

What made De Bary so very special? The extent of his list of major 'firsts' is quite astonishing and it is difficult to appreciate how each was in itself a

Fig. 3:2. Anton de Bary. "De Bary did not have a face for photography" according to Reess,[14] a biographer who knew him well, but his piercing blue eyes are clearly evident. (Courtesy of Hunt Institute for Botanical Documentation, Carnegie Mellon University, Pittsburgh, PA.)

major addition to the sum of contemporary knowledge (Box 3.1). Above all, he established a new methodology of research in mycology and also, as it became an independent discipline, in plant pathology. Old knowledge was no longer to be taken on trust; the new methodology required objective evidence borne out of original, detailed observations of the materials being studied and, most important of all, it required evidence that resulted from intervention and experimentation. This was the New Botany that De Bary helped to create. But from its many practical advances, each based on objective evidence, he was able also to make conceptual advances which linked apparently disparate fields of research, while raising fundamental questions that challenged his students for generations to come. De Bary was quick to

Box 3:1. Anton Heinrich De Bary (1831–88) and his major achievements

From his childhood in Frankfurt, De Bary loved plants but he graduated as a doctor of medicine. As part of the course (1849–53), he received lectures on medicinal plants and completed a dissertation on 'The sexual reproduction of plants'. He soon left medicine (1854) to become first an unpaid lecturer in botany at the University of Tübingen, then an extraordinary professor at the University of Freiburg-im-Breisgau (1855). In 1867 he moved on to Halle.

(a) Discoveries in mycology

1852: demonstrated organs and cycles of sexual reproduction in Phycomycetes (his name) and, also, sexuality in Ascomycetes (1862)
1866: separated Myxomycetes from Fungi, calling them Mycetozoa; his student, Rostafinski, published the first monograph on the group in 1874
1866: made the first reliable recognition of a nucleus in a fungus

(b) Discoveries in plant pathology

1853: in *Die Brandpilze*, argued from microscopical observations that disease arose only from infection by microbial parasites
1861 and 1863: demonstrated by experiment that potatoes became blighted after inoculation with spores of the fungus *Phytophthora infestans*
1865: culminating from studies of life cycles of rust fungi, demonstrated some rust fungi are heteroecious (have two unrelated hosts in their life cycle)
1866: with a prize-winning essay for the French Academy, finally demolished the theory that disease could arise by spontaneous generation
1886: reported that fungal secretions could disorganise host tissue, pointing to the role of enzymes in plant-pathogen interactions

(c) Conceptual advances

1866: published the first textbook of mycology, *Morphologie und Physiologie der Pilze, Flechten und Myxomyceten*, ever to go beyond simple listing and grouping; introduced the term saprophyte, distinguishing 'pure saprophytes' from 'facultative parasites' and 'obligate parasites'
1879: adopted and popularised the term 'symbiosis' coined by Frank in 1877, though even earlier (1869) he had been among the first to recognise lichens were products of a fungus and an alga

For more information see[02,03,04]

grasp the implications of Darwinism and evolution through natural selection which became an integral part of his thinking and of that of his students.[05]

He was special too in the way he managed to inspire and aggregate around him so many like-minded and brilliant young men: it was almost as though he had read a modern book on Management Science with its emphasis on the values of 'teamwork' and 'bonding'. Over many years he built and nurtured this network of men sympathetic to the New Botany. Not only was his laboratory open to visitors but, as we will see, De Bary took a personal interest in their studies. Keen to train them in his methods and to imbue them with his own ideas, he was nevertheless always willing to learn from them, something that was quite extraordinary in such an eminent professor. He was often fiercely critical, and sometimes intimidating, but he was never aloof like Sachs, his great contemporary. When his collaborators moved on, he rarely lost touch with them, corresponding regularly, advising them on their papers, which he often published in *Botanische Zeitung,* the journal of which from 1867 he was co-Editor, then Editor. They became part of an extensive virtual family of botanists.

By the meticulous study of fungal infection processes, which involved long hours of microscope study of thin sections cut from infected plant material, De Bary traced complete infection cycles for a number of plant pathogens. While the very best among the mycologists and pathologists who preceded him, such as Berkeley, had made their own observations of material collected from naturally infected plants in the field, De Bary went further. There were two novel strands to his method of research. First, he intervened; for example, he would sow spores on slides and on host tissues and record their development. Second, he experimented; typically this meant inoculating whole plants with their putative pathogens in order to test the true abilities of those micro-organisms to induce the symptoms of disease under natural conditions.

Most famously, in the early 1860s De Bary applied his methods to the study of potato blight. Incidence of the disease had rumbled on at low levels since its peak during the Great Famine of 1845–46 in Ireland, which had led to the death of over one million people and forced another 2 million to flee, mostly to North America. The cause of the blight was still unknown for the thread-like fungal hyphae seen growing out of infected leaves were regarded as evidence of a fungus which grew in tissue after its death, rather than of one which was the cause of death. While wild theories abounded, such as that leaves were choked by the exceptionally heavy rainfall, or the air was polluted by the emanations from the new railway locomotives, only a few rational voices argued that a fungus was the cause of the disease. Most notable among these were Berkeley and an ex-surgeon of the French army, Dr Jean Montagne, with whom he exchanged views. Berkeley had built a high reputation as a fungal taxonomist, a measure of which was that he was asked to classify the fungi brought back from Charles Darwin's voyage on the 'Beagle', but he was not an experimentalist. He offered no practical proof of his theoretical conclusions.

What De Bary did was to re-examine each stage in the life cycle of the fungus and to *manipulate* it as he extended his studies to include the germination of spores on both glass microscope slides and the surfaces of detached leaves. He saw how the germinating spores gave rise to a second generation of smaller 'zoospores' capable of swimming in water, thanks to their possession of two flexible, whip-like flagellae, before settling down at a spot favourable for invasion of the leaf. Next, he took a simple but nevertheless revolutionary step forward; he grew healthy potato plants in pots and inoculated half of them with the test fungus, taking care that the other half, the healthy 'controls', were not exposed to infection. He found, of course, that plants treated with the fungus developed symptoms characteristic of disease while the controls remained healthy. He noted that each blighted plant could give rise to millions of the primary spore type which could then either be splashed onto other leaves, by rainfall, or infect the tubers after being washed off the plant into the soil. By 1863 he had elucidated the complete life cycle and shown that this fungus was the sole cause of blight.[06]

The fungus was referred to as *Peronospora infestans* but when, thirteen years later, he re-visited potato blight at the request of the Royal Agricultural Society of England, it became clear to De Bary the fungus did not carry its asexual (conidio)spores like a *Peronospora* and that it should be put in its own genus, *Phytophthora.*[07] To this day the proper, Latin name, for the potato blight fungus, with its authority, is *Phytophthora infestans* De Bary.

The magnitude of De Bary's achievements are better appreciated when put in their temporal context. In 1860 Louis Pasteur had delivered a series of five lectures to the Academie des Sciences in Paris. In the two key ones he announced to the world what privately he had been convinced of for probably nine years. This was the fundamental principle that spontaneous generation, the creation of a living organism solely from inanimate matter, was impossible. Disease was caused by microbes, not generated *de novo* but produced by living parents and transmitted in the air, or by water, or by physical contact.[08] De Bary's studies of potato blight gave rational support to Pasteur's arguments. De Bary was aware of Pasteur, for in 1863 he wrote, "it [spontaneous generation] does not occur among my little fungi either". But was Pasteur aware of De Bary?

Pasteur was not a man who readily acknowledged the contribution of other scientists and, like most medical scientists of his age, he probably regarded plant diseases as being of no interest or importance. Unsurprisingly then, Pasteur never gave any indication that he was aware of either De Bary or his works. But was he hiding something? In 1861 Pasteur had won an essay competition organised by the Academie des Sciences, in which De Bary came second. If his attention had not been captured by the title of De Bary's essay, 'Researches on the development of some plant parasites', then it might be expected that Pasteur, a man who was more money-conscious than most, would have been aware of De Bary as a threat to his carrying off the first prize of 2,500 francs – a sum which compared favourably with the 2,000 francs that was Pasteur's annual research budget at the Ecole Normale. Whatever the case, Pasteur's essay was unanimously judged to be the best and De Bary had to be content with the second prize of 1,000 francs.

In his splendidly entertaining book, *The Advance of the Fungi*, EC Large wrote, "by 1866 De Bary had reached a point in the understanding of the causation of diseases, albeit in plants, that was *far ahead of the medical profession"*. It was not until 1884 that Robert Koch, the German bacteriologist and Frenchman Louis Pasteur's bitter rival, laid out his four Postulates. Still regarded as a turning point in microbiology and included today in virtually every student textbook of microbiology, the Postulates are the starting point for any investigation in medical microbiology. They require that to be accepted as the causal agent of disease:

1. the putative disease organism should be constantly present in diseased but not in healthy individuals of the host species (constant association)
2. the organism must be isolated and grown in pure culture
3. when the culture is inoculated into a healthy host it should reproduce symptoms characteristic of the disease
4. the organisms should be isolated from the inoculated individual and show, in culture, the same characteristics as the original isolate.

Over twenty years earlier, with *P. infestans*, De Bary had actually fulfilled each requirement for a plant pathogen, save being able to grow the micro-organism in pure culture, although others soon found how this could be done.

In passing, we should note that rusts, and a few other obligately biotrophic fungi, still prove the exception; they are accepted as the causal agents of disease in spite of never having been induced to produce their complete life cycle in culture.

To return to De Bary, in 1872, seven years before Harry visited him for the first time, he was persuaded to move to Strassburg. The move from the small and charming old provincial city of Halle was not an obvious one because, although modern Strasbourg is one of the most interesting and cosmopolitan cities in Europe – its old fortifications provide tourists with a charming contrast to the stark modern buildings of the European parliament – the Strassburg of 1872 was a city struggling to recover from siege and bombardment. Above all, it was struggling to come to terms with a change of identity – from being a French city to being a German city.

Situated in Alsace, on the borders of France and Germany, Strassburg in the mid-nineteenth century was basically a fortress at a crossing point of the river Rhine. It commanded both the Rhine valley and the great passes of the Vosges, a gateway through which, since Roman times, armies invading from the east might enter the plains of central France. In the late summer of 1871, during the second year of the Franco–Prussian war, the fortress and surrounding town had been subjected to a terrible battering by the advancing Prussian armies. Hundreds of homes were destroyed and the roof of the great medieval cathedral was badly damaged. Finally, on 27 September, the fortress was breached and the French defenders surrendered.[09] The siege had lasted 50 days. On his arrival in the city, Harry found it still clearly bore the scars of war. It was, moreover, a poor place in which to polish his German since everyone spoke the language with a strong French accent, an 'Alsatian twang'.

The fact that De Bary left Halle when he was settled and his career was flourishing to move to Strassburg tells us much about the political energy that was directing German science and education at this time. De Bary was offered the Rektorship of the newly named Kaiser-Wilhelm-University (Pasteur had been Professor of Chemistry in Strasbourg from 1849 to 1854, i.e. in its pre-war French days). In order to persuade him to move, the same inducements were laid before him as might be offered to a top scientist today – a better salary, a new building, new equipment, and new appointments. But it seems highly likely there was also some political arm-twisting, dressed-up as idealism. The University was firmly in the sights of the Iron Chancellor, Prince Otto von Bismark, the architect of German unity. Bismark wanted the university to demonstrate to the rest of the world, and to France in particular, the superiority of German education. Therefore, he appointed Baron von Roggenbach to re-define the university. Von Roggenbach was a man of great foresight; moreover, for the age, he possessed a remarkable political sensitivity. He gathered the brightest young professors from around Germany – at the age of 41, De Bary was by comparison an old man – but, in order to build goodwill with the people of Alsace, who still in their hearts regarded themselves as French, he appointed a significant number of Alsatians. Although De Bary was born and bred in Frankfurt, his family had Belgian-Walloon roots, so he might have been seen as a man of less than pure German stock and, thus, acceptable to those who still harboured anti-German feelings. Von Roggenbach even wanted to admit French as a language for teaching but that was overruled by Berlin. Despite the fact that the University swiftly gained the highest reputation for it scientific work, academics and students from Alsace did not fit seamlessly into the University, which never became the cultural melting pot of von Roggenbach's dreams.

De Bary was of small to medium height, slim and generally unremarkable appearance[10] except for his piercing blue eyes. Behind those eyes lay a man of great humour and warmth. Harry and the other young men who spent time with him were shown not only his scientific methods but, just as importantly, they were shown how a distinguished and demanding professor could unbend and relax with his staff and students, without losing dignity or authority. To be fair, the ability to relax and be sociable seems to have been something which De Bary, in his turn, had had to learn from others. He liked to tell a story against himself: during his early years (1855–66) in the small provincial university of Freiburg his dedication to work was not always appreciated by his colleagues. They told him he should leave the drinking of tea in his laboratory and get out and have a glass of wine with them in the evenings. Life was too short. They impressed on him that however hard he worked even he would never know everything. Wisely, he took their advice.

De Bary adopted what today would be called an 'open door' policy.[04,11] In Freiburg his workplace had been a rather large, low-ceilinged room, reached only by a steep, winding staircase, where he sat close to his students, readily available to them. Teaching them was the most important part of his life. Even in the magnificent Strassburg Institute, where as Rektor he was a man of great local importance and his fame was international, his working arrangements changed little; his office was a small room next to his laboratory and his research students. At intervals he would leave his own work for a short time and go through the laboratory, challenging students with his questions, always ready to feed to them ideas and 'leads', but always urging the them to do more work, to gather more evidence. This relatively private face contrasted with his public face; in the opinion of Harry, and of most others who heard De Bary speak in public, he was not a brilliant lecturer. In front of an audience he became shy and nervous; he was quietly spoken and each sentence was preceded by much nervous clearing of the throat while his fingers twiddled restlessly.

His private face was seen nowhere better than on the field trips which he loved so much. Looking back on those expeditions Harry remembered, "His close acquaintance with species and even local varieties of plants around Strassburg could only be known to those who have walked with him; and the delight of those walks in Alsace!".[05] Others recalled, "The stamina and speed of his walks, as well as the insensitivity to heat was not less admirable than his exact knowledge of plants".[11] Energetic until late in life, in spite of walking with a strong limp because one leg was shorter than the other as a result of a femur broken in 1863 when he was climbing to the so-called 'Rinken-hut' on the Feldberg, he was affable and witty. He loved to regale his companions with anecdotes about botany and botanists, while frequently stopping to answer questions or to point to some rare specimen or floral oddity. He had observing eyes that needed no glasses and missed nothing. "There are some young botanists these days" he proclaimed "who regard it as below their dignity to know the weed at the side of the footpath".

The last remark is significant because it shows De Bary, like other proponents of the New Botany, always taught that studies of internal form and function, of anatomy and physiology, should complement a proper knowledge of morphology and field botany and not replace the latter, as some contemporary critics would have had it.

De Bary clearly inspired great loyalty, none less than from Count Graf Solms von Laubach, who followed him Freiburg to Halle and finally on to Strassburg. Industrious and knowledgeable, Graf Solms was, nevertheless, at times a buffoon. He was the man whom the young students laughed with and sometimes at, but whom, one suspects, played a key role in keeping the laboratory happy and united. As a lecturer, he was at different times interesting and disorganised. "He tumbles over chairs, puts his fingers in mouth and goes through contortions that would make a cat laugh," the young American student, Will(iam) Farlow (Fig. 3:3), wrote to his mentor, Asa Gray, at Harvard. There were few pressures upon university staff and Graf Solms, who, in spite of his antics, many reckoned was a better lecturer than De Bary, attracted an audience of only six to his lectures on flowering plants. The number attending his lectures on Pilze [fungi] was not counted but he often delivered his lectures in a whisper so that De Bary, in the adjoining room, might not hear his mistakes.

Graf Solms had a fondness for beer and red wine. On one occasion, Farlow invited eight botanists to dinner, including De Bary and Graf Solms. "The supper lasted, for Strassburg, a short time, that is from half-past eight till a

Fig. 3:3. A young Will(iam) Farlow, protégé of Asa Gray and later leader of plant pathology at Harvard University and through North America. On receiving the picture Gray remarked "Thanks for your photo. Very nice! How your moustache sets you off!" (By permission, Archives of the Farlow Herbarium, Harvard University.)

little before twelve. The greater part of the company then retired to the Deutscher Kaiser to finish up with a few glasses (10 or 12) of beer."

None of Farlow's letters home gives a better feel for the relations between De Bary and his students than that dated Christmas Day 1873.[10]

> "Last night I went to De Bary's to see his tree. ... When I arrived I found the Norway spruce already lighted up. Instead of the hideous wax doll called the Christ child there was a bunch of gilded pine cones which Mrs De Bary had prepared for the occasion. There were four children, Wilhelm the oldest who was very sheepish and didn't say a word, August more lively, Marie, a rather too lively young lady six years old with a pug nose and little black eyes which are generally half strained out of her head by the two black pig tails on the back of her head, last night however her hair was flying in all directions, and little Hermann with very red cheeks and blue eyes. The presents were put on little

> tables near the tree. Hermann was perfectly absorbed and could not say a word all the evening. All the boys had a lot of soldiers. They seemed to have more toys than American children. Marie had only four dolls and a doll house besides a lot of other stuff and the older ones had books and games. The candies and cakes were a sight to behold and came from all parts of Germany from Berlin to Strassburg ... There was an indescribable thing which [I was told] was a "Pommern Gansebrust" and a great delicacy. It was a great goose skin sewed up at both ends and dripping with oil. It was full of uncooked smoked goosemeat and said to be delicious! The Pommern geese are very celebrated. There were also Erlangen Lebkuchen and Berlin gingerbread and two cans of condensed milk on his table. Mrs De Bary had amongst other things two bottles of arrack punch. At the table the company appeared much interested in my description of peanuts which they thought couldn't be very nice ... The other day De Bary had a great present from America. A box containing a swamp pitcher plant, some hickory nuts and twelve cranberries. As I said they were good to eat they cooked them and liked them very well."

Even if Strassburg itself was unattractive and had no society, the same could not be said of Anton De Bary, his family and his students.

In the 1870s and 1880s, Strassburg and Würtzburg were essential stops on the 'European Grand Tour' for aspiring botanists from Europe and North America. Sydney Vines, Isaac Bayley Balfour, Frederick Orpen Bower, Francis Darwin and Harry Marshall Ward were just some of young Britons who went and brought back the seeds from which the New Botany was to grow at home. The traditionalists looked upon the young men with amused tolerance. Writing in 1886 to Asa Gray, his old friend and the man who had encouraged Farlow to go to Germany, Sir Joseph Hooker said "I ... have thrown aside all idea of making headway with – any desire to keep up with even – heads of Chemico-botany, and Micro-phytology. I may content myself with a casual grin at young men calling themselves botanists, who know nothing of plants, but the 'innards' of a score or so." The stale, systematic botany of the traditionalists was eventually enlivened by the work of young men who knew the 'innards' of plants.[12] Hooker had undervalued the Young Turks.

Chapter four: From industrial Manchester to leafy Englefield Green

"A merchant came into the Principal's room accompanied by his son. After explaining that he wanted the boy to be admitted as a student, the Principal enquired which classes he would like his son to attend. The good merchant, jingling gold in his trousers' pocket with a proud consciousness he could afford it, replied, "Oh! I reckon he may tak' 'em all".

From the *Reminiscences* of WC Williamson, Professor of Natural History, Owens College, Manchester.[01]

A strong belief in advancement through self-improvement permeated Victorian society. It extended from the family to the city and to the nation, and its outcome was a hunger for higher education. Harry Marshall Ward was well placed to take advantage of the nation's appetite.

Every parent will understand Francis Marshall Ward's excitement when it was announced by the civic authorities in Nottingham that a University College was to open in the city, for not only did it offer the opportunity of employment for his son but also the possibility that Harry could return to live with, or near, his family. So, early in 1881, Francis wrote to his son in Ceylon urging him to apply for one of the professorships on offer. To Francis' disappointment, Harry was not so enthusiastic. He did not apply: the opportunity was missed and the posts were filled. On 30 June 1881,[02] he wrote to his father from Peradeniya, "I got the *Guardian* and read all about the professors – Fleming is an old friend of mine, and we used to meet every Saty [*sic*] evening at the Science Club meetings in Cambridge: he is an excellent fellow, and the authorities may be congratulated on his appointment: indeed, I think the selection all round is very fair, and probably the best men got the posts". Harry's decision not to apply proved fortuitous because the post had a hidden drawback. Ambrose Fleming, who was appointed professor of Physics, Mathematics and Mechanics, soon discovered the problem and resigned after only one year. Like most of the first professors, Fleming was expected to teach students from an impossibly wide range of educational backgrounds. On top of this, he was extremely frustrated by the university authorities' interference in his research.

Harry continued to his father, "You may have wondered, perhaps with some disappointment, that I did not go in for the berth; but there are many reasons, and I have to consider a good deal ere committing my valuable self to any scheme". Indeed, there must have been several factors weighing heavily on his mind at about this time. For example, he seems to have felt it was too early to leave Ceylon. Thus, throughout 1881, in response his mother's repeated urgings that he might return home, he explained that he felt he had not only a duty to his employers to see out his contract but, also, that fortune was currently favouring him for his project was at last going well. He was all too aware of the need to complete his researches in a way that would make them publishable and memorable. He recognised that this was his one opportunity to make a name for himself as a research scientist; if he were ignored upon his return to Britain then he would probably have to fall back on school teaching, the career which had beckoned in his South Kensington days. In addition, he probably foresaw the new college in Nottingham would not immediately enjoy a high academic reputation – growth was indeed relatively slow and it was not until 1948 that Nottingham achieved the full status of an independent university – so he may have wanted to leave himself free to try for a post in a longer established and more prestigious institution. What is likely to have been his overriding consideration, however, was that he was aware that his old mentor, William Thiselton-Dyer, was about to fix something for him in Professor Williamson's department at Owens College in Manchester. Dyer's scheme would have been doubly attractive because, firstly, Harry knew and respected Williamson from his days as a student at Owens in 1875/76 and, secondly, a job in Manchester would mean that he would be able to live in the same city as his fiancee, Selina. They could at last spend time together and plan for their long-intended wedding.

Harry had corresponded regularly with Dyer throughout his time in Ceylon and the latter was well aware that Harry's contract was soon to end. One evening at a Royal Society dinner, sitting next to Sir Henry Roscoe, Professor of Chemistry at Owens College, Dyer took the opportunity to promote Harry's interests by suggesting to Sir Henry that Ward would be a fitting recipient of a Bishop Berkeley Fellowship for original research at Owens. The arm twisting must have helped, although Harry still had to apply through the proper channels. He sent a written application from Ceylon and was duly elected to a Fellowship which carried a salary of £100. The *Minute Book of the Senate of Owens College* for October 1881 notes that four other applicants were shortlisted, including a geologist and a physiologist. Harry was chosen because "in his case the pecuniary aid afforded by the Fellowship would be of material assistance". "Mr Ward … expresses a strong and earnest desire to have an opportunity … and [in his application] refers especially to the phenomenon of dying and diseased cells, a subject of great importance to scientific agriculture." The Principal of Owens, JG Greenwood, reported, "the very important results already achieved by Mr Ward in Ceylon, in the domain of higher botany, led the Senate and the Council to make this appointment".

The Bishop Berkeley Fellowships were named after the churchman and philosopher, not the mycologist, but Fellows did not have to satisfy any religious qualifications or tests. Modelled on a similar scheme at Johns Hopkins University, Baltimore, the Fellowships were awarded for one year in the first instance and were designed to provide a stepping-stone for young academics to gain research experience.[03] Fellows were expected to devote themselves full-time to their special area of study and at the end of the year to demonstrate their progress either by thesis, lecture or some other appropriate means. What the Fellowships offered the College was a means to build its academic credibility through fostering the researches of the brightest young men of the day.

Owens College had been established in 1851 in a private house on Quay Street, with five professors, two lecturers and 62 students, but it had grown rapidly and moved in 1873 to new buildings on Oxford Road, the present site of the University of Manchester. Its early history is typical of many colleges and institutions which at this time were part of the first expansion of higher education since the middle ages – an expansion that was driven by the growth of industry and its need for scientific and technical expertise.

The industrial revolution spawned many great cities in Britain but few were bigger or dirtier than Manchester (Fig. 4:1). Its population in 1801 was a mere 93,000. By 1871 it had more than quadrupled to 475,000. Men and women were drawn to the cities by the lure of employment and cheap housing. Out of the grime and exploitation there grew, however, the phenomenon of 'civic pride'. The wealthy merchant middle classes of these cities were fiercely competitive, each wanting for their own city the finest buildings and the noblest institutions. As good Victorians, they also had an unshakeable belief in the virtue of 'self-improvement'. Among such men in Manchester was Solomon Behrens, uncle of Harry's friend Edward (chapter 2). Solomon and a group of wealthy businessmen raised over £200,000 to finance the move of Owens College to the Oxford Road site. Their ambitions for Manchester extended to having libraries and a fully-fledged university. They were largely successful: opening the Free Library in 1872, Sir James Stephen, Professor of Modern History at Cambridge summed up the mood of the times, "At Cambridge, as everywhere else, where educated men are brought together, Manchester is a name of deep and even of awful significance; for here is the metropolis of that Titanic industry, on the continued success of which England has pledged her station and her authority among the nations of the world".[01]

Owens College may have been the starting point where the fight for a university in Manchester was concerned but students taking degree courses still had to register through the University of London, and teaching courses at Owens had to follow the rules and regulations set down by London. A step forward was taken in 1880 when the Victoria University was created as a federal institution that joined Owens with similar colleges in Leeds and Liverpool, and ambitions were finally realised in 1903 when Manchester University was granted its own separate Royal Charter (Leeds and Liverpool Universities received their own charters at the same time).[04,05]

The success and growth of Owens can be largely attributed to the quality of the staff appointed in its early years and none was more influential than WC Williamson (Fig. 4:2). He was one of the first professors appointed but

Fig. 4:1. A contemporary view of smoky Manchester. Small boats had been able to navigate the 36 miles inland from the sea to Manchester since 1740 but in 1882 a group of prominent local businessmen decided to build the Manchester Ship Canal. Opened in 1894, the Canal allowed the largest ships to berth in Manchester. (From Punch 7 October 1882.)

unlike others who soon moved on and away, such as the distinguished chemist Edward Frankland, Williamson remained at Owens for the rest of his working life. His appointment as Professor of Natural History was initially half-time, for which he received a salary of £150 and a portion of the students' fees. He had no staff to support him in those early years but student numbers were small and teaching made few demands upon his time. Much of the lecturing was done in the evenings, so he was able without too much difficulty to retain the post of surgeon at the Manchester Ear Institute while keeping-up some sort of medical practice in Fallowfield (then just a village on the outskirts of Manchester, rather than an integral part of the city as it is now).

Although enrolment in the first classes matched expectations, with city merchants well represented at evening classes, numbers of day students had

Fig. 4:2. William Crawford Williamson FRS, palaeobotanist and Professor of Botany, Owens College, Manchester. (By permission, the Royal Society.)

dipped so low by 1857 that the future of the college was threatened. The professors, led by Edward Frankland, argued that in spite of Owen's being established to instruct and improve young persons (of the male sex) in such branches of learning *and science* as are usually taught in the English Universities, the training received by most students was too classical, fitted only for those students who sought BA or MA degrees; the ethos resembled too closely that of Oxford and Cambridge. While recognising that the Principal, the Rev. AJ Scott, formerly Professor of English Literature at University College, London, was a man of the highest intellectual culture, the professors complained that he was not sufficiently sympathetic to the sciences or in tune with the needs of local employers. Their criticisms – plus external pressure led by the *Manchester Guardian*: "this college, which eight years ago it was hoped would form the nucleus of a Manchester university, is a mortifying failure" – persuaded Scott to resign. He was replaced by JG Greenwood who, although he was a Greek scholar, was a man much more in tune with the needs of the city and much more to the liking of the science professors. Led by chemistry, where the energetic Henry Roscoe was quickly brought in to replace the disenchanted and departed Edward Frankland, student numbers surged upwards.

The greater numbers and the imminent move to Oxford Road, which for the first time provided Williamson with a teaching laboratory, persuaded him to resign his medical posts and concentrate on his academic career. In 1872 a professor of Geology, Boyd Dawkins, had been appointed, enabling Williamson to relinquish his responsibility for teaching that subject, and in 1880 Milnes Marshall was appointed to a chair of Zoology, enabling Williamson to relinquish that subject too. At last he was Professor of Vegetable Physiology and Botany, and nothing else. His salary was now £250 plus the usual share of the fees. After publication of Darwin's *The Origin of Species* in 1859, in which so many of the arguments were potentially verifiable through the fossil record, Williamson had become recognised nationally as one of the few authorities who could bring fresh objective evidence to the greatest intellectual debate of the age. His workload increased enormously and in 1878 he was allowed to share his burden of teaching by appointing Marcus Hartog as his personal Demonstrator and Assistant Lecturer.

Williamson was a leader in the foundation of the discipline of palaeobotany – the interpretation of the fossil record of plants – but he was essentially a self-taught, old fashioned naturalist. Since his boyhood in Scarborough, where his father was curator of the town's museum, he had assembled a unique collection of rocks and fossils which formed the basis of his early studies. He welcomed the appointments of Hartog and, later, of Ward not just because they took work off his shoulders but because these younger men brought with them a more modern and detailed knowledge of plant anatomy. He was as happy to learn from them as he was to teach them about the fossil record of plant evolution: their relationships were mutually beneficial.

Harry's Berkeley Fellowship, begun in 1881, was extended beyond the first year, finally expiring on 1 May 1883 when he replaced Hartog as Demonstrator and Assistant Lecturer. Commenting to the College's Senate about Harry's publications during tenure of the Fellowship, Williamson wrote, the "high value of his [Ward's] contributions to botanical science is abundantly recognised by those who are competent to judge of their worth". (As a mark of that distinction, Harry had been in 1883 elected a Fellow of Christ's, his old Cambridge College.) Apparently it was not held against Harry that, apart from completing the publication of work begun in Ceylon, his research at Owens had not directly addressed the phenomenon of diseased and dying cells, which he had mentioned in his job application. Instead, he published on 'The sexuality of fungi'[06] and on 'An aquatic

Myxomycete.'[07] The latter paper includes a passage that reveals his growing ingenuity as an experimentalist and, at the same time, gives a nice feeling for the conditions he experienced in Manchester. "... in order to obtain information as to the feasibility of employing the electric light in the botanical laboratory, so that demonstrations of the assimilation of green plants might be made independently of the sunlight, of which we obtain little in Manchester ..." during winter and spring, he grew plants under Swann's incandescent lamps.

Self-reliance and ingenuity were the essential ingredients for survival in grimy, gritty Manchester. This was Harry's first significant appointment carrying teaching duties and he met the new challenge successfully. Indeed, when he resigned his Assistant Lectureship at Christmas 1885 his students "showed their respect and esteem by asking him to give them an extra lecture over and above the course". One could not imagine the same thing happening today, however brilliant the lecturer! Harry had been equally popular with his colleagues who, at the time of his leaving, held a complimentary dinner in his honour. As well as his research and teaching and the seminars he gave at Owens, Harry was, with characteristic energy, a contributor to meetings of the Manchester Literary and Philosophical Society (Manchester Lit. & Phil). He gave lectures not just about his studies of coffee rust[08] but also about the infection processes of *Saprolegnia ferax*, a fungus which exploits lesions in the skin of salmon infected by a pathogenic bacterium.[09] The disease was highly topical for its recent greatly increased frequency in rivers from the Scottish Borders to North Wales had both alarmed land owners and sportsmen and caused Her Majesty's Inspectors of Salmon Fisheries to devote the whole of one of their annual reports to the subject. The Inspectors' report, authored by Mordechai Cooke, argued that hunting of otters had reduced their stocks and so, in a Darwinian sense, had allowed unfit salmon to proliferate. Outraged by that conclusion, TH Huxley was prompted to make one of his few ventures into mycology and, from experiment, he came to the conclusion that the fungus spread in fresh water from dead flies to the salmon. Unfortunately, Huxley should have stayed clear of mycology for his conclusion was wrong. Equally unfortunately, Harry told the Manchester Lit. & Phil. that his own results confirmed Huxley's, though he did redeem himself by describing, with typical attention to detail and accuracy, the infection processes of *S. ferax* as it (saprotrophically) colonised salmon tissues. He also distinguished himself by describing methods by which he had obtained in large numbers the fungus' oospores, the resting spores which are so important in the epidemiology of the disease because they allow the fungus to survive in the absence of the host. A study of their ecological role would have been richly rewarding for Harry, but he did not have time for everything!

Personal matters were for once intruding upon Harry's science. The greater financial security given by an Assistant Lectureship, meagre as it was, persuaded him to end his long engagement and in the summer of 1883 he and Selina Mary were finally married (Fig. 4:3). Of Selina, called 'Lina' within

Fig. 4:3. A young Selina Marshall Ward (nee Kingdon), photographed on the Isle of Wight, possibly during an early holiday. (By permission, the Royal Botanic Gardens, Kew.)

the family, we know little directly though we may confidently infer that she was both patient and understanding. Mostly, she bore calmly the long wait for marriage although early in their relationship she had shown an occasional flash of spirit, as in September 1877, just prior to Harry's second year as a student in Cambridge. There is no record of her specific complaints but, clearly, they were sufficient to shake him out of any complaisance about their relationship into which he might have fallen. He replied[10] to her letter(s):

"My darling,

I have had a long walk, thought, done some work, again pondered and hesitated, and have at last determined to write another letter. At first I determined to tear up the one I wrote this morning (on receipt of yours), but, again, I will also send it, for there is on it the stamp of my temper at the time I read yours ——— You can thus judge at the effect your cruel words had, though even these I curbed to some extent. I can only hope – feel that you never meant those words to reach me: What shall I do? How can I be sure that you trust and love me as I do you? Who will guarantee your faith of happiness in my love, if you so lightly put such bitter words on paper? I do not know what to say to please you: I only know that in this way have begun our little unhappinesses in the past (when you thought you could trust me and I knew you could not), I only fear their aggrandisement in the future.

'tis true, darling, I couldn't pray, and seek a text, and then go to sleep happy, when any mistakes arise; but I can and do most earnestly try to account for the error.

Come, I have done. I wish we were together and then these wretched little griefs should be soon cured by kisses. However, I can claim one very dear recompense for my unhappiness ——— a real, long, loving, trusting letter, such as my pet can write when she chooses.

I do not say I have committed no fault – I only know I am ignorant wherein it exists. However, say you forgive me unless you intend to adhere to words which have caused me more wretchedness than any you have ever used to me. If there is nothing in it, to you – say so at once: Our course is then too plain ~

With love

Harry.

Be careful with me, darling, you can never know how much I feel all this."

On the back of his letter he copied the words of 'Mein Liebchen' (My Darling) a popular romantic balled of the day.

Mein Liebchen, mein liebchen!
canst thou remember yet,
those halcyon days – those happy days,
    when we so often met;
the sunshine – in springtime,
thro' flowery glade and lane,
how oft' we'd roam we two alone,
Aah when will they come again?
    the years have pas'd, the love is young,
We fondly plighted then,
    Oh, come my darling say to me,
that melody again –
    That fav'rite song – our own love song,
        Mein liebchen!

Two more verses followed. Selina was won over; they remained engaged to be married.

For almost three years of the engagement, Harry was in faraway Ceylon (1880–82). By his own admission, he wrote more often to his mother and sisters than to his fiancee. To his mother he wrote, "I suppose I am guilty in a way of neglecting Lina – i.e. not having written letters enough and all that; but you see I am not much of a fellow at those things, and girls are too fragile and funny for rough chaps like me to understand them".[11]

Early in his stay in Ceylon, he had broached the subject of marriage, though a sense of duty, almost pity, and timeliness, rather than passion, seems to have been his motivating force. He consulted both his fiancee *and* his mother, apparently *simultaneously*, about the possibility that Lina might come to Ceylon, where they would be married. His letter to Lina is lost, but to his mother he wrote,

> "There can be little doubt that if ever I marry at all it ought to be pretty soon – and there can be no doubt that it also ought to be Lina. Now it happens that my present position is such that I can support a wife in a modest way if that 'blessing' should be forthcoming. Moreover, I think I know of another party who is of similar opinions, and who would willingly come out to me here if she were asked to do so. You know her, and you know me (rather, eh?): what shall you say if I send for her in the course of 6 months or so? Well, I know you have some doubts about us: but let me impress upon you a fact that is established by my observation, and is not a creation of a heated lover's brain or any such nonsense – any display of temper … you may have noticed lately has been caused by me alone. There is no mistake about it, but she is a plucky constant little thing to put up with the neglect she has so silently borne for some time. It seems at times as if my work drove every passion & thought of others away from me, and tho I used to think such enthusiasm and devotion the highest virtue, I am beginning to believe that it is a greater thing to be able to work for others as well as for one's self."[12]

Luckily for Harry, the two women in his life shared the same opinion. Thus, he could reply[11] to his mother's letter,

> "I come to the conclusion you do – viz:– it is not worth while to incur such an expense as we should thereby incur, and the possible joys are dearly bought by people like us, who are so poor in all worldly materials. But you will be glad to know that she too – brave and patient as she is – has refused to come, perceiving what a useless expense such a proceeding would involve, and prefers to wait until I have finished my work out here".

As often in the course of Harry's life, a lack of money had dictated the course of events. Lina stayed where she was, in cold, dirty, smoky Manchester.

Through most or all of these years in which her fiance was in Ceylon, Lina lived with her mother, her father having gone to Texas, USA (chapter 2). The reasons why Francis Kingdon left remain obscure but differences between his religious beliefs and those of his wife may have led to unbearable tension. Whereas he came from a family where Anglican clergymen abounded, and there was even a cousin who was Jesuit priest, she belonged to a new Fundamentalist sect, the Plymouth Brethren (chapter 5). Whatever the reasons for his exile, the report of his death in 1881 was a severe blow and could only have intensified Selina's sense of loneliness and isolation. Writing to his own mother shortly after the death, Harry reflected, "Poor lass, she has indeed lost a friend in her father; for he alone understood her, and the most touching trust and love existed between them".[13]

Lina must have had nagging worries about the future of her relationship with Harry, but it is more likely they were that he would never marry than that he would find another girl. She would have known in her heart what Harry admitted to his mother:

> "The fact is, I am already wedded to my work, and for some years have played the part of a bigamist – so to speak – If I don't love Lina, it is not probable that I shall ever love any girl, and you must admit it would be cruel and cowardly if

> I failed to try and make her happy who has trusted me so long and so well. Do therefore be very careful in all your letters, and avoid anything that might hurt her. Your copy of letters makes me fear she is very unhappy without me, and I feel very wretched sometimes when thinking how poorly I have returned her gentle treatment of me – you see I cannot tell even you all that has passed. How she has forgiven my negligence over and over again! And you must own she has been very plucky to resist this chance of coming out."[11]

Lina's patience was eventually rewarded. Their wedding took place at the Parish Church of St Paul, Charlton in south-east London, close to the new home of her mother, widowed in 1881, and of her brother, Zachary, a naval engineer who was probably working in the nearby shipyards. The service was very much a Kingdon-family affair, being conducted by her cousin the Reverend James Durrant Kingdon, Headmaster of Sutton-Valence School, Staplehurst. (Sadly the church was a victim of one the bombs that rained on London's dock area in the Second World War.) The newly-weds returned to Manchester where they had found a house in Withington, a suburb just to the south of Owens College.

Lina's married life could not have been easy. She could never have been under any misapprehension that she did not have to share Harry with his work but only four months after their wedding she had to share his attentions with his family. On 4 December 1883, the happiness of the young couple was cruelly interrupted by the death of Harry's 'Darling mother', Hannah Mary Ward, at the age of 50. The effect on Harry is hard to underestimate. As his letters show, even within the bounds of a close-knit family, their relationship had been exceptional. She had at different times been his nurse, counselor, friend and even student. Although his father, Francis, was undoubtedly well supported in a practical sense by his three daughters, he and they needed emotional support from Harry, the oldest child and only boy. His youngest sister, Nelly, was only 17 years old and especially needed her big brother's love and attentions. Harry must have had to spend much of his time away in Nottingham, consoling his father and sisters.

The roller-coaster year for Harry and Selina's emotions had not finished yet, but the next event was a happy one. Two months after the death of Mrs Ward senior, Mrs Ward junior became pregnant. On 12 October 1884, Winifred was born in Charlton. As often happened with first pregnancies in Victorian times, the expectant mother stayed with her own mother for the birth. Just a year later, on 6 November, 1885, Frank was born. This time, Selina was confident enough to give birth in the family home, 14 Heaton Road, Withington.

Now, the ambition of a young man is never sharper than when he is newly married: pride and the need to support a wife and, often children too, are the natural driving forces. In the spring of 1884, less than a year after his marriage, Harry would then have been overjoyed to learn that his old friend Isaac Bayley Balfour was vacating the Regius Chair of Botany in Glasgow to take up the Sherardian Chair in Oxford. Here was the chance of a prestigious professorship and a better and more secure income. With little hesitation he

Box 4:1. Support for Harry Marshall Ward from members of the botanical 'Establishment'

(a) For the Regius Chair of Botany, Glasgow, 1884–85

Writing to Dyer from Cockburn's Hotel in Edinburgh on 19 February he recounted,[14]

"I went to Glasgow yesterday with introduction to McKendrick (Physiol) and Jack – saw them, and the Principal, as well as Young (Nat History). Reception seemed favourable; but all agree that I must have influence in Edinburgh, and advised my seeing Dickson (Professor of Botany). Hence I came here. ... Am just off [home] to Manchester, Dickson is away from home; and I have just missed Bayley Balfour, who came to London on Saty night."

Nevertheless, he managed to back his application with a list of eminent supporters: JG Greenwood and WC Williamson (respectively, Principal and Professor of Botany, Owens College), Michael Foster FRS (Professor of Physiology, University of Cambridge and Secretary to the Royal Society), Julius Sachs (Wurzburg), W Carruthers (British Museum, Natural History), Rev. Miles Joseph Berkeley FRS (the respected mycologist), Henry Trimen (his old friend and colleague) and his life-long mentors Sydney Vines and William Thiselton-Dyer.[15]

(b) For a Fellowship of the Linnean Society, May 1886

Harry was sponsored by JG Barker (Secretary, Botanical Society of the British Isles), Daniel Oliver (Keeper of the Herbarium at Kew and Professor of Botany, University College, London), JD Hooker and WT Thiselton-Dyer. Ironically, these were the men who just a year earlier made the final decision to give the Glasgow chair to FO Bower.

decided to apply for the post. He sensibly decided to undertake some lobbying in the right quarters, visiting both Glasgow and Edinburgh and assembling an impressive list of referees (Box 4:1a).

He must have understood that others would be interested in the Glasgow chair and that his candidature might be unsuccessful but he could have had no inkling of the machinations that would surround the appointment and his own, ultimate, rejection. Added to a final feeling of rejection, the whole affair must have left questions in his mind as to the level of support he had been given by those he trusted and relied upon, especially Dyer and Vines.

We have to try to piece together what happened from the subjective accounts given after the event by various players in the drama. Objective evidence is missing. Thiselton-Dyer, as always a key figure in Ward's life, wrote in 1912, "I got so deep in that Business that I thought it fair to all those concerned to burn the file about it at Kew".[16] Government records were destroyed soon after the main events in 1885, probably also to avoid any future embarrassments.

Why was government involved?; simply because at that time appointments to Regius Chairs had to be approved by the Crown – in effect, by the Home Secretary. The complications started because Balfour wished to resign his post at Glasgow in March of 1884 but he was not needed at Oxford until the start of the new academic year in October. Thus Glasgow University invited Balfour to remain in post during the summer term and, most importantly, it did not officially accept his resignation until the end of the summer term, so delaying the appointment of his successor. Meanwhile, three weeks into the summer term, the Professor of Anatomy at Glasgow (Cleland) received a letter from Professor William McNab of Dublin saying that he had been appointed by the Home Secretary, Sir William Harcourt, to the Regius Chair of Botany.[17,18]

Quite understandably, the Senate of the University of Glasgow was at first surprised and then, when it had time to gather its collective thoughts, outraged at having the decision about such an important decision made for it. Despite objections from some of the Senate 'old guard' who resented too many English appointments to Scottish chairs, and the fact that in his favour McNab was a Scot, it seems Harry had been the favoured candidate. There was a deadlock which stretched through the latter part of 1884 and into the early months of 1885. The government's resolute support for McNab may well have been based on political at least as much as academic considerations. Prime Minster Gladstone's second administration was struggling, as it turned out unsuccessfully, to find a coherent policy that would lead to Home Rule for Ireland and it needed all the political goodwill it could muster. The removal of McNab, an unpopular Scot, from Dublin would have opened the way for the government to curry political favour in Ireland by the appointment of an Irish professor. Unfortunately for the government, the implacable antipathy of the Glasgow Senate towards McNab only hardened and then the botanical establishment turned against him.

Although McNab's academic standing was not as high as that of Ward, he would, even without the government's intervention, have been a serious candidate for the Regius Chair. An Edinburgh graduate whose father and grandfather had held positions at the Royal Botanic Garden there, he had been Professor of Natural History at the Royal Agricultural College, Cirencester where he succeeded Thistelton-Dyer. Two years later, in March 1872, he again followed in Dyer's footsteps, becoming Professor of Botany at the Royal College of Science in Dublin. McNab was a prolific author on plant morphology and physiology, a hard worker, and very ambitious. Indeed, he had applied for the chair in Glasgow when it was previously on offer in 1879; on that occasion having lost out to Bayley Balfour. The reason the botanical establishment turned against the well qualified McNab was that during his years in Dublin he had developed a reputation for being difficult, demanding and abrasive. For example, when on 1 April 1878 the Department of Science and Art in Ireland took over the administration of grants which maintained the Royal Dublin Society's Botanical Gardens at Glasnevin, McNab seized the opportunity to obtain better facilities at the Botanical Gardens for his own research and teaching. Among other things, he wanted "the exclusive service of one of the gardeners, whom … he desired to obtain specimens for him in the Gardens".[19,20] The long-serving, able and well-respected Director of the Botanical Gardens, David Moore, complained to Dyer that McNab, with others, was plotting to force him into unwilling retirement, whereupon he [McNab] could become Director. Moore was convinced that McNab could hardly wait to get his hands on the keys of the conservatories and glass-houses. In the event, David Moore died in 1879 after a short illness. To McNab's chagrin, Frederick Moore (David's son) was rapidly appointed Curator at Glasnevin and McNab was given only the junior role of Scientific Superintendent. It was for such reasons that McNab's candidature for the Glasgow chair lost support.

The deadlock was broken, not in Glasgow but at Kew, by four wise men. Sir Joseph Hooker, Thiselton-Dyer, Daniel Oliver and Mr JG Barker (see Box 4:1b) put their heads together and in April 1885 they summoned into their presence Frederick Orpen Bower. Bower, an old classmate of Ward (chapter two) and a man whose career had followed the familiar, well-worn track to the German laboratories of Sachs and De Bary, was teaching at South Kensington and doing research in the Jodrell Laboratory at Kew. He had applied for the Sherardian Chair at Oxford and come second to Balfour. Having failed at Oxford he was probably looking towards Cambridge where the ageing professor, CC Babington, could not live much longer. Bower had not considered Glasgow so it was a complete shock to him when one day, peacefully working at Kew, he was summoned by Hooker. In Bower's words, "Sir Joseph said he supposed I knew about the vacant chair in Glasgow. I said, Yes, but that I had decided not to apply for it. He then told me that there had been a hitch in the appointment, and that a new candidature was desirable. That he considered it to be my duty to apply, and that if I consented the four of them would support me. I asked for twenty-four hours to think the matter over, though I knew that whatever their reasons I was bound to obey." The following morning Bower told Sir Joseph he would comply. In his memoirs[21] he gives the impression that from thereon everything proceeded rapidly and automatically, though others have pointed to the fact that he had to apply and go through, at least, some token selection procedure. Whatever, the four wise men of Kew had achieved a political fix; they had found a candidate acceptable to both government and the Senate of Glasgow University.

Why was Harry's candidature was passed over? All the reasons are not known but critical factors must have been Hooker's attitude and Dyer's stance. Joseph Hooker was probably the man with the final word since he was the senior botanist in Britain (he was also the son of Sir William Hooker who had occupied the Glasgow chair for many years), and Bower, but not Ward, was known to him. In contrast, Dyer knew both Bower and Ward but, from his personal knowledge of Harry, he had doubts as to whether Harry had the right drive and personality to make a success of what was a difficult job. Dyer judged that Bower was the tougher, more self-confident man. So, on 18 March 1885, Dyer gave his full support to Bower, writing "I have already given testimonials to Mr Ward and Mr [Walter] Gardiner (another original applicant) ... But I have no hesitation, as you have since determined on being a candidate, in emphatically expressing my opinion that the Government could not obtain a candidate in any way more qualified. ... Professor Huxley, the Dean of the Normal School of Science, has repeatedly expressed to me his satisfaction and pleasure in the way you have acquitted yourself in your duties, and, as you know, he is not a man to bestow his commendation lightly."[22] Vines, then a Reader at Cambridge, also supplied a testimonial for Bower.[23] He ranked Ward, Gardiner and Bower as about equal in attainments, but gave Bower the edge because of his South Kensington experience and being more in tune with the New Botany! It is clear today that the latter was a serious misjudgement on the part of Vines, because by the 1920s the young 'Botanical Bolsheviks', led by one of the new leaders of British Botany, Arthur Tansley, were fiercely critical of the 'Scottish morphologists', Bower and Balfour, for having abandoned physiology, ecology and genetics.[24]

As with the earlier post at Nottingham, so with the chair at Glasgow, it may have been Harry's good fortune that it was not he who was appointed. Bower's first years at Glasgow were not happy ones. Aged only 29, he was expected to give elementary botanical teaching to 300 or more medical students, accommodation was in his own words 'meagre', he had to use his own money to purchase microscopes from the outgoing Balfour and within weeks he came to realise the Botanic Gardens, of which he was automatically joint-Director, were £50,000 or more in debt. Later in life he admitted "I tried to get away again but was pulled up by Thiselton-Dyer and told to stick to my job and await (or cause) developments. The isolation in Glasgow was terrible to me after 10 years in Cambridge, Germany and London. There was no-one to talk my botany in Glasgow, only Dickson in Edinboro' and Trail in Aberdeen and neither of them was modern."[18] Bower did indeed 'stick to it' and he did 'make things happen', staying happily in Glasgow through the remainder of his working life. Bower and Ward remained friends and later in life Harry was probably able to realise that the loss of the Glasgow chair was not an unmitigated disaster; he must have taken much satisfaction from the fact it was he, not Bower, who finally succeeded Babington at Cambridge.

Harry's disappointment about the Glasgow chair was in the shorter term somewhat ameliorated by his appointment in the same year (1885) to the new Chair of Botany within the Forestry Branch of the Royal Indian Engineering College at Cooper's Hill (Fig. 4:4). The College had been founded in 1871 to provide increased numbers, and to improve the standards, of engineers entering the Public Works Department in India. Approval for the College was pushed through parliament by The Secretary of State for India, the Duke of

Fig. 4:4. The Royal Indian Engineering College, Cooper's Hill, from the *Illustrated London News*, 25 November 1871. "Every student before proceeding to India will be required to furnish the President of the college satisfactory evidence of his competancy in … rifle drill … and in riding" (From Cooper's Hill Calendar, 1888/89, p.125.)

Argyle, apparently against the wishes of both the Prime Minister, Mr Gladstone, and the Government of India. Predictably, with such political opponents it did not last long and was dis-established, finally, in 1906. The *Cooper's Hill Magazine* of June 1904 wryly summed up the history of the College:

> CHAPTER I 1870
>
> The Secretary of State: "I propose to establish the College"
> The government of India: "We do not like the idea."
> The S. of S.: "Oh! Don't you? Then consider it established."
>
> CHAPTER II 1903
>
> S.of S.: "I propose to dis-establish the College."
> The G. of I.: "Don't do that it has been a great success."
> The S. of S.: "Oh! Has it? Then consider it dis-established."

In 1885 training in Forestry was added to that in Engineering, again the motive being the need to meet the requirements of the Indian civil service for large numbers of well trained young men. In Forestry, 10–15 students were admitted each year to the 22-month long course, which was followed by 4 months of intensive practical training in British or German forests. Later the course was extended to three years with nine months in Germany. On closure of the College in 1906, all activities in forestry were transferred to Oxford.

Harry's appointment came, happily, when the College was in an expanding phase and his ten years there coincided with its most confident period.

Cooper's Hill (now the Runnymede campus of Brunel University) was a most beautiful place, celebrated by poets:

> "On Cooper's Hill eternal wreaths shall glow
> While lasts the mountain or while Thames should flow"
> (Alexander Pope).

The main building, made famous in two classic films, 'The Belles of St Trinians' and 'The French Mistress', overlooks the Thames and has distant views of Windsor Castle. Nearby is the site at which King John signed the Magna Carta in 1215 and, long after Harry's time, the John F Kennedy Memorial was built. The last fatal duel in England was fought on the estate in 1852. What greater contrast can there have been then with dirty, noisy, overcrowded Manchester? Cooper's Hill must have seemed to Harry an idyllic setting in which to bring up a young family. With the comparative riches of a professorial salary, the Wards were able to move into a large, solid and eminently respectable Victorian home, 'The Laurels', at nearby Englefield Green (Fig. 4:5).

The ten years Harry spent at Cooper's Hill and 'The Laurels' were not just chronologically at the center of his adult life. They were at the heart of his family life for they were years in which he saw his young children grow (Fig. 4:6) and their personalities emerge; they were at the core of his professional life for they were years in which his scientific reputation grew and he became accepted as part of the botanical establishment of the day. It was for him a period of enormous personal energy and achievement. While he would happily cut himself off from the world, often spending continuous days and nights recording what he saw under his microscope, he still retained from his youth a strong sense of fun. He could unbend with his children, making up games for them and sharing with them his wonderful world of biology.

Winifred, in an unpublished biography of her brother Frank, whom she idolised, paints a revealing picture of these years when the children were young.[25] "We were allowed a great deal of freedom, … we were surrounded by woods, hills and streams, to say nothing of the Thames itself, and the garden was large and old-fashioned, with plenty of scope for running wild." "We were taken to London at intervals for shopping or to visit [mother's] relations, and we found it bewildering but exciting, with its noisy clip-clopping hansom cabs, buses and drays, all drawn through the cobbled streets by horses."

In contrast to the warmth of her feelings for Frank, Winifred's descriptions of her father are cooler, but by no means unloving.

> "When we were small he would sometimes get annoyed with us and punish us gently if he felt we needed it. But the only thing that ever seemed to make him really angry was any suspicion that we were not being strictly truthful; then he would thunder. Exaggeration or fanciful statements about factual matters he would correct, as likely to prevent us from recognising the truth when we saw it."

Fig. 4:5. "A view of 'The Laurels', the family home at Englefield Green, showing spectroscopic experiments being carried out by father (circa 1893)". Winifred's comments, inscribed many years later on the back of the photograph, show how Harry took his work home. (By permission, the Royal Botanic Gardens, Kew.)

> "He was a marvellous teller of tales of all kinds. But he was always careful that they should not become confused in our minds with the facts surrounding us and with what we could, and were led to observe."

In keeping with the mood of the times, Harry was always keen to 'improve' as well as to amuse his children.[26] Thus, he performed at home simple chemical experiments to amaze and delight them; it is not recorded whether he borrowed reagents and equipment from his laboratory or purchased the same from one of the many firms which specialised in supplying Victorian households. Imagination, fantasy and plain fun were important in the time he spent with them. He used his well-honed drawing skills and knowledge of botany to create imaginary plants with funny names based on objects and events with

Fig. 4:6. Selina with her son Frank, photographed at 'The Laurels'.
(By permission, the Royal Botanic Gardens, Kew.)

which they were familiar, such as *Bicyclia multipunctata*, and *Wriggleaboutis agooddealia*; two more are shown in Fig. 4:7. He put the 'nonsense' poems of Edward Lear to music, taught them songs from the popular light operas of Gilbert and Sullivan, and even made up his own verses. When he found that the biscuit box contained only a particularly unattractive sort called 'Ginger Nuts', he proclaimed:

"I do perceive – I thought I should –
Some biscuits that are made of wood;

A dark brown colour, like the oak!
Now am I not a funny bloke?"

The children remembered on another occasion, when the cat tried to steal food from the table, he rhymed:

"Kitty came – ah well I mind her,
Licked a plate of mutton broth,
And departing, left behind her
Footprints on the table cloth!"

Music, of course, featured large in Harry's leisure time. While accompanying himself on the piano, he would sing to them, in his fine baritone voice, songs that he had composed.

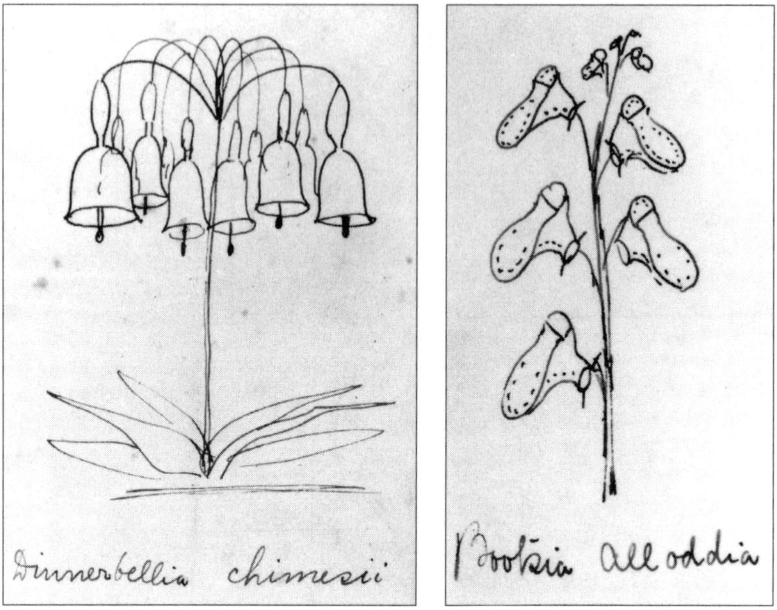

Fig. 4:7. Cartoons drawn by Harry for his children, *Dinerbellia chimesii* and *Bootsia alloddia.* (By permission, the Royal Society.)

The adult lives of Harry's two children could not have been more different from each other, apart from their common and enduring lack of money. Winifred remained unmarried throughout her 94 years, taking care of her ageing mother and later working as a speech therapist, always living in London (Fig. 4:8). By contrast, Frank's life was much more eventful. Married twice, divorced once, with two children by his second wife, Frank Kingdon-Ward (he and his sister preferred to be known by this surname) was perhaps the last in the great tradition of European plant-hunters who since the 18th century had travelled fearlessly to India, Nepal and China to bring rare plants back to the herbaria and horticulturalists of Europe and North America. Most notable among his introductions was the blue poppy (*Meconopsis betonicifolia*), a plant with which his name will always be associated.

> "Everyone who knew Frank recognised in him a quality of freshness and eagerness which he never lost," wrote Winifred, though she might equally well have been writing about her father in his younger days. In Winifred's view, Frank was notable for his modesty and integrity. "The first was perhaps as complete as one could find in anyone, and was due probably partly to upbringing and partly to inheritance. Our father was just such a man."

A vital part of Frank's inheritance from his father was an early awareness of the world beyond Britain, in particular of the variety and richness of plant life in the tropics. Thus,

Fig. 4:8. Winifred photographed in her teens, after the family had moved from Englefield Green to Cambridge. (By permission, Cambridgeshire Collection, Cambridge Central Library.)

> "When … Frank was a small boy something happened which, he maintained afterwards, sowed a seed from which the whole of his later career was to grow. Our father was walking up and down the lawn of our house, in conversation with one of his cronies who had recently come from India, while Frank played about near them. 'There are places up the Brahmaputra,' said Father's visitor, 'where no white man has ever been.' This single sentence stood out … it was then that a picture of the mysterious valleys below the roof of the world was implanted in Frank's mind."

In Charles Lyte's fascinating biography of Frank Kingdon-Ward (1989), he writes that the intellectual gifts Harry gave his children were not matched by practical, tangible gifts. Harry's austerity meant that they enjoyed few luxuries or comforts, and none of the dolls' houses and rocking horses that furnished the nurseries of their wealthier friends. Lyte says that the only time they ever had a Christmas tree was when they stole one from a public park – though, if Harry knew of this, it seems highly unlikely he would have approved, given his high moral principles. Whatever the details, the capture of the tree was just one of Frank's many secret pranks and escapades. But why was Harry austere, perhaps penny-pinching? To some extent, the habit was ingrained in his childhood and was a reaction to the problems he had suffered as a result of his own father's financial near-ruin. But added to this was the fact that, although the professorship at Cooper's Hill was only modestly paid, Harry and Selina had ambitious views on the way their children should be educated. They employed governesses to teach the children at home when they were little. Like a growing number of middle-class Victorian parents, they sent their oldest son away to boarding school. So, in 1893, when Frank was less than eight years old, he had to leave home for Colet Court, the Preparatory School for St Paul's; later, in 1898, he progressed on to the famous London public school itself. Quite understandably, Harry wanted his son to have the proper uninterrupted education of which he had been deprived. He and Selina achieved this, but at considerable financial and emotional cost to themselves. Frank's behaviour at St Paul's was to cause his parents regular bouts of anxiety but he was not allowed to leave the school for it had been decided that Frank was going to go to Cambridge. He was going to follow in his father's footsteps.

Chapter five: Work will enable us to forget

"His profession was botany and his outlook that of the scientist, and his whole life was devoted to a search for ultimate truth."

Winifred Kingdon-Ward writing about her father, Harry Marshall Ward, in *The Flower Chief.*[01]

Cooper's Hill was, in spite of the charm of its location, not entirely to Harry's liking. Writing after Harry's death, Dyer[02] recollected, "the utilitarian atmosphere in which he [Harry] found himself was not very congenial to him. But he had at any rate at last some sort of adequate position and a laboratory to work in, and here he remained – not, I think, unhappily – for ten years." Working in an establishment dedicated to forestry, located on a site in which 25 of the 100 acres were woodland, Harry naturally developed his existing interest in trees. He took his students each week during the summer months to visit the Arboretum at Kew. At Kew there was also an opportunity to view the magnificent collection of over 1,000 timber products, which had been presented by the Indian Government in 1878, plus the collection representing economic botany which had been rescued from the India Museum when the Great Exhibition closed. The Indian specimens must have brought back happy memories of his adventures in the jungles of Ceylon. Less than a day's journey away were 50,000 or more acres of beech woods on the Chiltern hills, while, closer to Cooper's Hill, visits were made to the nearby Windsor Forest where oak and Scots Pine dominated 10,000 acres of woodland. Reflecting his new preoccupation with forestry, he wrote two books on the subject, *The Oak* published in1892 and *Timber and Some of its Diseases* published in 1897, and he also edited a new translation from German into English of Robert Hartig's groundbreaking *Textbook on the Diseases of Trees* (1894).

A Professorship at the Royal Indian Engineering College may not have given Harry quite the kudos he desired, but his personal reputation grew steadily as more and more of the established, senior scientists came to know him and his works. Meetings of the British Association for the Advancement of Science (BA) were one way through which he became integrated into the scientific 'Establishment'. He had been a member of the BA since his student days in 1879. He read his first paper in 1883 at the Southport meeting; entitled 'On some cell contents', it was an account of fatty bodies in the cells of coffee. As a mark of his progress, he had by the 1887 meeting in Manchester become deeply involved in the organisation of the Biology section of the meeting – of which Botany was a sub-section (see chapter 7). The botanical 'Establishment' of the day was relatively small and the names of many of those who organised parts of the meeting will already be familiar. Thus, among the meeting's Vice-Presidents were WC Williamson (the motivating force behind the invitation from Manchester to the BA) and Thiselton-Dyer, but there was also Asa Gray from Harvard, that most eminent of all North American botanists, with whom Harry was previously unacquainted. Harry was co-opted to be one of six secretaries for the Biology part of the meeting. He was also invited to join a committee reporting on the establishment of a Botanical Station at Peradeniya, Ceylon. The committee received and approved, no doubt with Harry's strong support, a report from Henry Trimen suggesting that one of the bungalows in the Garden should be converted to a properly equipped research laboratory. Sitting on the committee were men Harry knew already; there was Bayley-Balfour, Carruthers, Thiselton-Dyer, Bower and Michael Foster. Also there was Marcus Hartog, who had preceded Harry as Williamson's demonstrator at Owens College and who was by now Professor of Natural History at Queen's College, Cork. At the meeting, Harry was a joint author of a paper 'On polar bodies' with Professors Weissman, Lankester and Krause, and Messrs Gardiner, Sedgwick, Carnoy and Hartig, all eminent scientists who were coming to know and respect his abilities.

Among other notable papers read in the botanical sessions was one by a rare husband and wife team, Professor and Mrs Percy F Frankland, on 'Studies on some new microorganisms obtained from air'. It was probably Harry's, first acquaintance with Percy, who was a chemist, and his wife Grace (nee Toynbee), who was a talented microbiologist, but he was to get to know Percy very well when a few years later they jointly produced for the Royal Society a monumental report on the microbiology of drinking waters (chapter 6).

Meetings of the BA were traditionally designed to appeal to the well-informed public as well as to schoolteachers and to the large number of amateur, gentlemen scientists of the day. Attendance at the Manchester meeting of 1887 was typical. Professional scientists were relatively few among the huge total of 3,838 delegates, but their number did include most of the 92 'foreigners' who were recorded. Williamson was approaching the end of his career and was determined to spare no efforts to make the botanical gathering a brilliant success – a grand finale (Fig. 5:1). Accordingly, he set about inviting the most eminent botanists of the day. Apart from Asa Gray, there was De Bary from Strassburg, Ferdinand Cohn from Breslau and The Marquis De Saporta from Aix. All these, plus Mrs Gray and De Saporta's son, stayed as house-guests of the Williamsons. Sachs also was invited but replied that he was too old and too ill to travel, though, even in his prime, he had rarely left Würtzburg. An invitation was sent to another great German botanist, Pringsheim, in Berlin but just two days before the meeting Williamson was told that he had sent an invitation to the wrong Professor Pringsheim – a chemist, not the famous botanist. Telegrams were

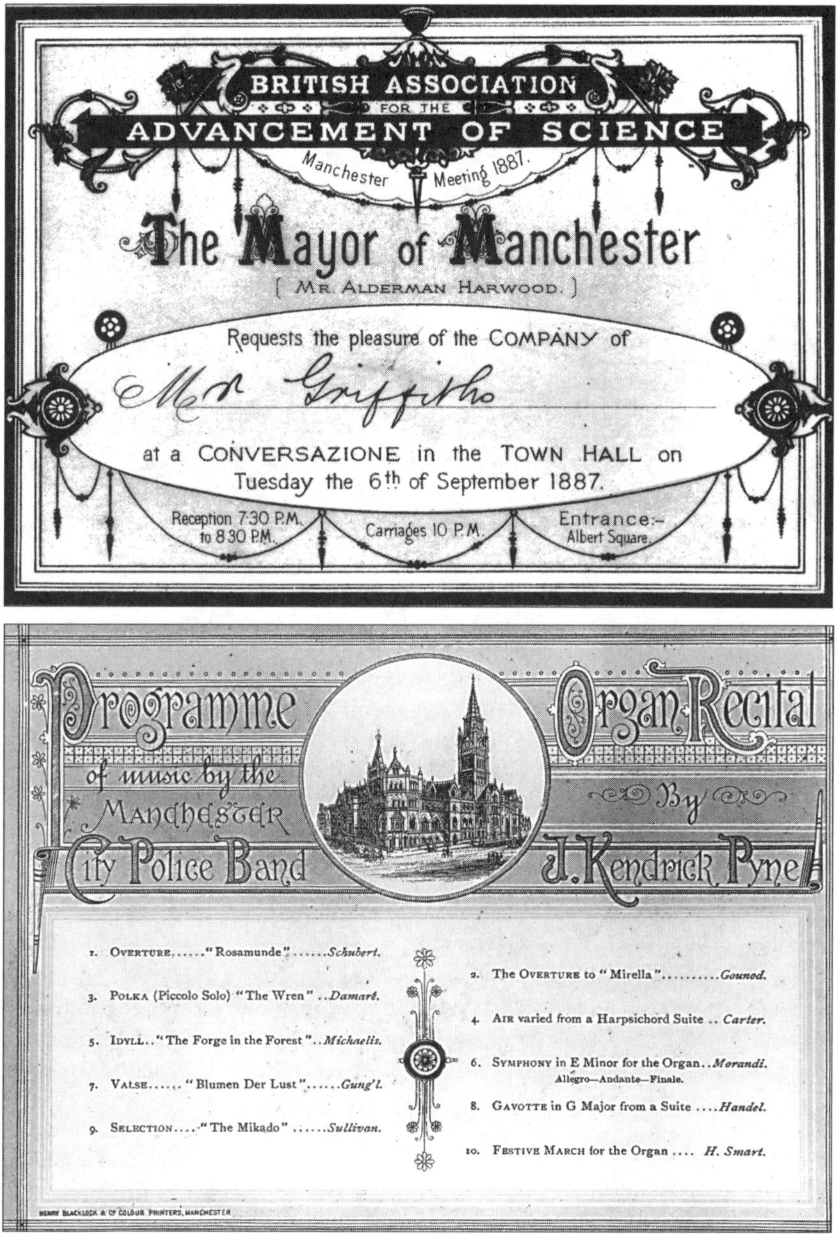

Fig. 5:1. Among the excursions, receptions and entertainments arranged for those attending the annual meeting of the British Association for the Advancement of Science held in 1887 in Manchester was a Conversazione in the town hall. The reverse of the invitation (below) lists the programme of popular light music. (MS. Dep. B.A.A.S. 172, by permission The Bodleian Library, University of Oxford.)

Fig. 5:2. Distinguished botanists attending the BA meeting in Manchester. Harry Marshall Ward (third from left) sits at the feet of Anton De Bary. To the right of De Bary is WC Williamson. (From Sparrow, FK, 1978, Mycologia 70, 222–52, by permission of The Mycological Society of America.)

hastily sent to the right Professor Pringsheim who, luckily, was free to travel and took his place among the distinguished guests. In his memoirs,[03] Williamson recalled, "… at other meetings the Sunday afternoons were often dull, and [we] tried to prevent this for our friends, by issuing invitations for a small assembly of naturalists in our own house and garden. Fortunately, the afternoon was fine. … One enthusiastic friend said to my wife that, 'most of the distinguished botanists of Europe and America were in the garden, and not one but who had seen something growing he never saw before'. Seldom in the 19th century can there have been such an assembly of the greatest botanists.

To his great satisfaction, Harry was amongst the guests invited to the Sunday afternoon party and, when the commemorative photograph was taken (Fig. 5:2), he placed himself appropriately at the feet of Anton De Bary. As well as renewing his acquaintance with the great man, he was also able to spend time with another old friend from his Strassburg days, Graf Solms. The latter's career had flourished and by this time he held a professorship in Berlin. Evidently, however, the good man was still a valued social organiser. Williamson wrote, "In order to avoid *contretemps*, our good friend Graf zu Solms Laubach, who knew most of the assembled Germans, came each morning to overhaul my wife's invitation list; to his kind help we are indebted for pleasant intercourse with our many guests".

New Botany in Britain, or 'The Cause' as Harry liked to call it, took a major step forward in 1887 when a new journal, the *Annals of Botany*, was launched by three of its leading figures, Vines, Bayley Balfour and Farlow.

In order to encourage North American involvement, Asa Gray had been invited to be one of the three founding Editors but he had declined, recommending his young Harvard colleague, Farlow. The three were "assisted by other botanists",[04] as they had been in the earlier planning phase when they had consulted with many of their contemporaries, including Harry Marshall Ward, about the need for a new journal which would specialise in publishing papers centered on the results of investigations conducted by experiment. Not surprisingly, they had received widespread encouragement and, perhaps more surprisingly, guarantees of the financial support needed to launch such a venture. The journal was to be for all English-speaking people, not just the English, and, although Farlow was at pains to scale down projections of North American sales from 100 to 20, the journal flourished and is still flourishing.

The first edition appeared in August 1887 and its very first paper was by Marshall Ward H & Dunlop J, 'On some points in the histology and physiology of fruits and seeds of *Rhamnus*'. Appropriately for a journal with such origins, the first book reviewed was *Lectures on the Physiology of Plants*, by Sachs; it had been translated into English by none other than H Marshall Ward! The anonymous reviewer wrote,

> "The days of hack-translators, – at least as far as science is concerned, – are, we trust, numbered. Every one now-a-days expects that a scientific book shall be translated by some one conversant with its subject matter. In this respect Professor Sachs' latest book is fortunate, and in welcoming the translation we must record our satisfaction at seeing the name of Professor Marshall Ward on the title page as that of translator. He has done his work solidly and well, and has produced a readable and trustworthy English version. And if here and there a Germanism remains, we must remember how penetrating and insidious an essence this same German is …":

this seems ironic in view of the debt the New Botany owed to Germany. Harry never became an Editor of the *Annals* but he was one of its most frequent contributors, as well as a regular referee.

The foundation of the *Annals* in 1887 coincided with the start of correspondence between Harry and the American plant pathologist, Will Farlow, with whom he eventually struck a firm friendship. They must have long been aware of each other for both were students and admirers of De Bary, and they shared many interests, but the first letter from Harry to Farlow, dated 10 September 1897,[05] is cautiously reserved, starting, "It is very kind of you to give me the opportunity to know you better". It then touches briefly on the subjects of papers which the two men were about to publish – 'Black Knot' and 'Onion Smut' from Farlow, 'Tubercles of Leguminosae' from Ward. But the letter ends more chattily, with Harry remarking, "I have lately returned from a visit to Robert Hartig in Munich, I had the good fortune to see much of tree-killing Hymenomycetes" [members of the toadstool family]. Such friendly tones were soon to be the norm in the letters between the two men as each recognised in the other a similar sharp intellect and a passion for plant pathology. They exchanged papers and gossip. In February 1888, Harry offered his sympathy for the great loss Farlow must be feeling at the death of Asa Gray (whom Farlow was to succeed at Harvard as Professor of Cryptogamic Botany). Gray was just one of a number of leading botanists who died about this time, the greatest shock for Harry being the death of his idol, Anton De Bary. He wrote to Farlow, "… as for the irreparable loss at Strassburg, I don't know what to say or think … Poor De Bary, he seemed so full of health and energy when here in September [at the BA meeting in Manchester], and we all thought him good for 20 years longer."[06] Ward and Farlow finally met in June 1892, at a hotel in London where Farlow was staying while visiting Britain. They were to renew their acquaintance five years later when Harry went with the British Association to Toronto.

Nature was still a relatively young journal in 1888 and was not pre-eminent as it is today, but it was nevertheless a feather in Harry's cap that it was he who was chosen by the journal to write De Bary's obituary. This was a task to which he gave the highest priority; he reported, "they telegraphed for it just as I was off to dinner at Schlick's. … I sat up all night at it, and although it was hurried work to get in this week, and had to be short, I hope you [Thiselton-Dyer] will like it."[07]

Harry Marshall Ward's name will forever be associated with coffee rust and Ceylon but probably his most important contribution to plant pathology was made while he was at Cooper's Hill, some eight years after returning from the tropics. In one paper,[08] 'A Lily Disease', published in the *Annals* in 1888, he helped to open a whole new approach to botany – he became a founding father of physiological plant pathology. It is a discipline which sought, and still seeks, to understand and ultimately to control disease by focussing on the functional interaction between plant and pathogen. Its aim is not just to understand the chemistry of the interaction, and to relate chemistry to structure, but to understand too the physics of the three-way interaction between plant, microbe and environment. As it evolved it sought to provide a rational basis for disease control strategies which at the time centered around fungicidal chemicals but which were soon to include the emerging discipline of plant breeding.

At the time, of course, the way ahead was not so clear. What Harry was attempting to do in his study of an obscure and commercially unimportant disease of lily was to show that one particular fungus contained within its mycelium, and released from the tips of the hyphae comprising that mycelium, something which broke down the structure of plant cell walls opposing the growth of those tips. In assessing the magnitude of his discoveries it is important to put them in the context of our current knowledge, while remembering how little was known about plant biochemistry at the time. For example, most people today will have heard of enzymes, if only in connection with biological washing powders, but in 1888 there was no agreed, well-defined concept of what was an enzyme, even among scientists. The action of such substances had been studied in chemistry laboratories across Europe since the early eighteenth century. Thus, in 1785, Irvine reported the liquefaction of starch by aqueous extracts of sprouted barley and, in 1833, Payen and Persoz were able to show the responsible agent, which they called 'diastase', was inactivated by heat and could be concentrated by precipitation in alcohol. But it was not until 1867 that the word 'enzyme' (meaning *in yeast*) was coined by Kuhne.[09]

Knowledge of the biochemistry of plant cell walls was equally elementary and confused. We know now that each plant cell is surrounded by a relatively rigid, non-living, cell wall whose internal skeleton is comprised principally of chains of the polysaccharide cellulose, closely packed into fibres, themselves embedded within an amorphous matrix of mixed polysaccharides. There may be air spaces between neighboring cells, the largest and functionally most important being those which facilitate gaseous diffusion between mesophyll cells in leaves, but, in the main, walls of juxtaposed cells are cemented together by pectic polysaccharides, which are amorphous rather than fibrillar. Largely through enzymic attack, but with physical force playing some part, most fungi create for themselves points of entry into or between cells of the epidermis, the outer layer that protects leaves, (young) stems, fruits and roots. (In contrast, rust fungi often bypass the epidermis, growing through the stomatal pore and invading first the mesophyll cells of the interior of the leaf.) Invading fungi secrete, in a sequence that is critical for success, a cocktail of different sorts of enzymes, each of which specifically degrades a different type of cell wall polysaccharide. Additional enzymes are produced to break down those lipid-rich polymers that comprise the waxy cuticle covering the epidermis of leaves and other above-ground organs. (In contrast, bacteria or viruses can infect plant tissue only through a natural opening, such as a stomatal pore, or a wound.)

Broadly, two patterns of fungal attack are recognised. Pathogens like rusts cause minimum disruption to their host because polymer-degrading enzymes are released by them in a targeted pattern, for example cellulase enzymes (see Box 5:1) are strictly localised around the tips of the hyphae growing into a cell wall. The fungus thus avoids causing host cell death by employing this conservative attack. These 'biotrophic' pathogens are at an advantage because host cells are enabled to live longer and, thereby, to continue synthesising nutrients. By contrast, 'necrotrophic' pathogens like *Botrytis* are less controlled, releasing enough enzymes to disrupt tissue rapidly and in advance of the spreading mycelium. Most critical in the latter mode of attack are pectic enzymes for these attack the main polymer of the middle lamella. As this region is degraded, tissues progressively lose structure and there is the breakage of any plasmodesmata, the cytoplasmic strands that connect adjacent cells through cell walls. As a consequence there is disruption of both the structure of tissues and any functions co-ordinated between cells by the plasmodesmata. In advanced cases, and especially where fleshy fruits and vegetables are invaded, the disease or condition is known as a 'soft rot' (Fig. 5:3).

There are no sharp or absolute boundaries between cellulosic and pectic polysaccharides (see Box 5:1), although each type of polysaccharide in the

Box 5:1. Polysaccharides from plant cell walls and their breakdown by enzymes

Cellulases and pectinases are polysaccharidases, enzymes that break down polysaccharides, respectively, cellulose and pectin.

Cellulose in plant cell walls has a complex, crystalline structure and complete breakdown to its constituent monomer, glucose, requires the action of three different types of cellulase. Thus, the term 'cellulase' denotes a complex of enzymes rather than a single enzyme type.

There exists also a family of 'pectic' enzymes, reflecting the family of 'pectic substances' which are their substrates. All pectic substances have in common a backbone or chain predominantly made of galacturonic acid units (monomers of sugar acid) which are joined by end-to-end by alpha links between carbon atoms 1 and 4 of adjacent units. If more than six per cent of the units in a chain have their carboxyl (–COOH) side group on carbon atom 5 in methoxylated (–CH_2OH) form, the polymer is known as pectin, or polymethyl galacturonic acid. Some enzymes are specific for the unmethylated, 'pectic' polysaccharide, others for the methylated form. Maceration occurs when chains are shortened by enzymes that break the alpha 1–4 linkages. The most common method of cleavage is by hydrolysis, where water is added, –H to one side, –OH to the other. The most effective cleavage, in terms of maceration, is by *endo*-enzymes, ones which attack at random along the chain length; thus, if it occurs in middle of the chain, the first cleavage can reduce chain length by 50%.

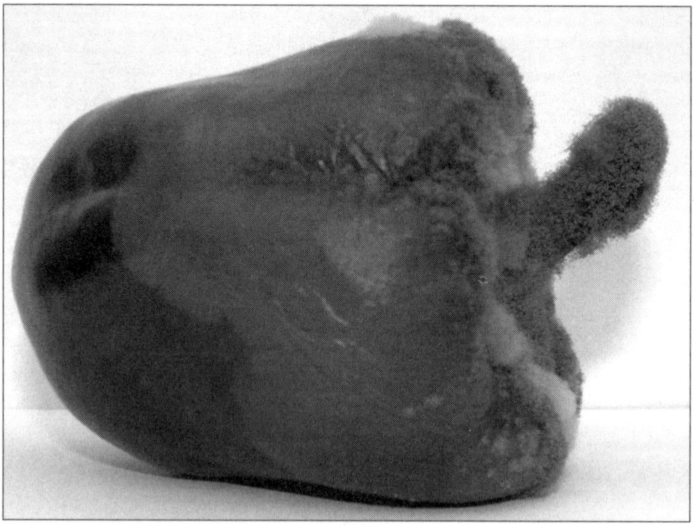

Fig. 5:3. Soft-rot of a *Capsicum* fruit caused by *Botrytis cinerea*. Infection occurs typically through a wound, as in the cut stalk.

plant cell wall is typically more concentrated in one region than in another. Add to this the fact that the middle lamella, particularly in cereals and other monocotyledons, also contains hemicellulosic polysaccharides (a diverse group in which each chain may be made from more than one type of monosaccharide) and we can appreciate why it is difficult even today to determine by microscopy alone which polysaccharidases are active in an infected tissue. We can appreciate also why it is nearly impossible, even with hindsight, to be absolutely certain which wall components and enzymes were being studied by Harry and his contemporaries.

To Anton De Bary, as for so much else, must go the credit for publishing in 1886 the first paper concerning the causes of a fungal soft rot.[10] Working with carrot roots infected by the fungus *Sclerotinia libertiana* [*Sclerotinia*, or *Peziza* as it was known to Harry, is the generic name given to the sexual, or perfect, stage of the fungus known in its asexual, or imperfect, stage as *Botrytis*], De Bary reported that there was partial or total dissolution of certain constituents of cell walls. Most importantly, from infected tissues, he could extract sap that would macerate healthy tissue. Activity of the extract was lost if the extract was heated – suggesting to us that it contained enzymes, since heating above 40°C inactivates most enzymes – but he merely referred to the extract as a 'ferment'.

Harry was aware of De Bary's work but in the Introduction to his paper on the lily disease, he makes no mention of it. This was probably because he regarded De Bary's fungus as a saprophyte, an organism feeding off the dead tissues of others, and, therefore, not the same organism as the parasite which he was studying. There was at the time a more rigid distinction between parasites and saprophytes than is recognised today; nowadays we believe many saprophytes (correctly, and hereafter, called 'saprotrophs') can be opportunistic parasites under certain environmental conditions or where host resistance is otherwise compromised. However, intriguingly and perhaps edging in the direction of a more modern viewpoint, towards the end of his characteristically long paper Harry wrote of De Bary's fungus as being "a form physiologically midway between the ordinary *saprophytic* Pezizas, and my parasitic one; in other words, De Bary's fungus is in the process of *being educated to parasitic habits*" [HMW's italics]. He took this position because De Bary's fungus was able to invade living tissues only after it had been established for some time in dead ones.

Harry explained how his interest in soft rot arose: "… during the summer of 1886 I had frequent opportunities of noticing … orange-brown and buff specks which appeared on the stems, pedicels, leaves, and buds of the white lily (*Lilium candidum*) growing in my garden. … Occasionally I found small tufts of a fungus springing from the spots." At first he regarded the tufts as belonging to a saprotroph and this view was confirmed when, in 1887, he cut sections through lesions and "found what looked suspiciously like a definite mycelium in the dead tissues". However, he was forced to change his view in the following wet summer of 1888. Starting from the orange and buff spots on the buds, with which he was already familiar, he clearly saw a "severe epidemic disease" developing in which the spots turned to greyish mouldy patches and the buds soon shrivelled and died. By mid-July, every leaf on the lower parts of the plants was decayed.

Using the classical methods he had first employed so successfully in Ceylon, that is by searching through a series of thin sections cut from tissue at different stages of infection, he was able to ascertain the disease was caused by *Botrytis*, a fungus which had previously been regarded as a saprotroph. At a mid-stage of infection he saw tissues just below the buff spots were utterly destroyed but in the surrounding green parts of the leaf there were no hyphae. From the densely packed hyphae in the epidermis of one surface, however, thin branches descended almost vertically to emerge at the opposite epidermis. Most significantly, he saw the hyphae in young spots were confined to the plant's cell walls, "in the swollen substance of which they are growing and branching in all directions". Similar swellings, which turned brown and granular, were seen in walls of epidermal and stomatal guard cells. Proof of the parasitic nature of *Botrytis* was completed when inoculation of lily tissues with spores of the fungus led to hyphal invasion and comparable swelling of host cell walls.

After his exhaustive time-series study, Harry concentrated his attentions on the germination of fungal spores on thin glass coverslips, 20 × 20 × 0.001mm, and the subsequent development of hyphae. From observations that hyphal tips exuded a 'deliquescent substance', particularly at branching points and where neighboring hyphae fused, he came to the remarkably prescient conclusion that "the continuous forward growth of the apex of any hypha takes place … [when]… the ferment-substance at the apex keeps the cellulose of the hypha at that place in a soft, extensible condition, and the pressure from behind stretches it and drives the tip forward". Each hypha was truly a thread of death.

Harry was naïve in assuming the fungal cell wall was comprised only of cellulose but he had anticipated by some 100 years the theory that hyphal extension occurs when turgor pressure, generated inside the hypha, exceeds a value sufficient to stretch irreversibly the relatively thin hyphal wall at tips and branch points. He had observed and correctly offered a physiological explanation for the fusion of hyphae, by dissolution of their walls juxtaposed at points of contact, which is essential both for the building of complex fungal structures and for sexual reproduction.

The next step was to try to determine how plant cell walls were pierced by the tips of the fungal hyphae. There were no precedents for this type of investigation and in the hands of lesser men it may have failed. Harry succeeded because, as well as possessing the energy and patience needed to follow the classic methods of investigation through microscopy, he possessed the originality and flair (and dexterity) of a great experimentalist.

Using the fine tip of a sterilised needle, one spore was lifted carefully from a parent colony and sown in a single drop of sterile distilled water onto a sterilised glass coverslip. Into this drop was placed a thin section of either an ovary from a young lily bud, or of the central scales of a lily bulb cut with a perfectly clean razor and washed to remove most starch grains (such scales remained contaminant-free for over one week in distilled water). Within three days mycelium developed. As growth of tips of hyphae proceeded, the thin cellulose walls of the bulb became swollen and evidently softened. Tips of hyphae continued their growth in the plane of what would be the middle lamella. Hyphae occasionally bored through from cell to cell, but did not directly attack remaining starch grains, nuclei or other cell contents.

Harry believed hyphal tips exuded a 'ferment drop' and he argued those drops should, in isolation from the fungus, have solvent action on cellulose [growth in the middle lamella suggests there may also have been action on pectic substances]. The next step in the investigation was to test this hypothesis. This required Harry to go much further than anyone before him had done – it required him to make detailed studies of the properties of the extract. Again, there were no precedents, so he had to find more ways of keeping media and equipment sterile. His basic technique was to sterilise Pasteur's liquid nutrient medium (he boiled it for 10–15 mins each day for several days) and then, under sterile conditions, to add to it glass coverslips whose microscopic examination showed him bore *Botrytis* spores, and no other microbes. In two weeks the fungus grew across the surface of the medium. Then he carried out four series of experiments.

In the first, he filtered-off the liquid medium and placed into it pieces of lily tissue. After 40h, examination showed the cut ends of all pieces, but especially of the peduncle (main flower stalk), were swollen and gelatinised, unless the filtrate had been previously boiled in which case the cut ends were still clean and unswollen. In the second series, he dealt with the mycelium. It was filtered from the growth medium and ground in a mortar to extract its liquid contents. Half was then boiled and lily tissue was added to both boiled and unboiled extracts. As before, the tissue in the unboiled, but not in the boiled, extract lost its shape. In the third series, he followed the same steps as in Series 2, except half the unboiled extract was filtered before use. It still had macerating activity.

The fourth series of experiments was rather different. This time he took mycelium from six-week old cultures, extracted its contents, filtered the extract and added it to four to five times its bulk of alcohol. After standing overnight, the flocculent white masses were extracted, dried and added to lily pedicel (stalk of the flowerlet). After 8h at 30°C, the middle lamella of the parenchyma cells was destroyed and the cells were isolated as if they had been boiled. Their cellulose walls were also swollen. Similar sections lying in boiled solution or in water showed no change.

In short, he had shown that when *Botrytis* is grown in liquid medium it synthesises, and will secrete into the medium, something(s) that will cause the swelling of cell walls and maceration of uninfected tissues. There is every indication that those things were enzymes, because the action of mycelial extracts was heat-labile and withstood precipitation in alcohol. Incidentally, in precipitating the enzyme in alcohol he had anticipated what was to become a standard biochemical method for the purification of enzymes.

Harry's view was that these four series of experiments related strictly to the fungus' saprophytic action, since it occurred on dead lily tissues. To assess its parasitic action, he continued his experimental studies by sowing spores, suspended in distilled water, onto the surface of young lily buds or leaves. Within 48h the germ tubes were found "in the solid cellulose-substance of the cell walls of the epidermis", their tips having attached themselves to the surface of the cuticle and then dissolved their way in. Whether or not infection progressed, he concluded, depended on whether or not the infected area lost water and shriveled sufficiently fast to trap the young mycelium. He thus recognised the importance of the environment for he realised that, in a humid environment, the ferment could diffuse "and prepare a path of destruction necessary for the rapid progress of the hyphae, the growth of which is of course dependent on the presence of water". His perspective, which recognised the importance of the environment to the outcome of infection, was, at the very least, uncommon. It was another indicator of his exceptional mind.

By careful observation of natural infections followed by experiments in the laboratory, De Bary and, to a much greater extent, Harry had laid the foundations of our understanding of the enzymology of cell penetration and tissue maceration, though neither had used the word 'enzyme'. In spite of its importance the subject area advanced exceedingly slowly after them, partly because it became sidetracked by discussions about oxalic acid. There was much confusion as to whether oxalic acid, which is known to be produced by a number of fungi, could either alone or in combination with a ferment be the primary cause of plant cell death. The truth is that oxalic acid is important in a very few host plant-fungal pathogen combinations but, unfortunately, it was some of these which had attracted the attention of the early researchers. Generally, however, understanding of enzymes and of their target substrates improved gradually and by 1910 LR Jones, in a review of bacterial soft rots,[11] was able to propose in what we would regard as modern terms that the bacterium *Bacillus carotovorus* secreted an enzyme – which he called 'pectinase' – with the power to dissolve the middle lamellae of certain vegetables well in advance of the invasion of the bacterium. Notably, too, Jones was amongst the first to make the direct connection between the action of such enzymes and host cell death. Harry's principal intellectual beneficiary was however a young man from the Scottish Borders, William Brown.

After a spell as a temporary lecturer under Harry's old friend Isaac Bayley-Balfour, at Edinburgh, Brown moved in 1912 to Imperial College, London, where in 1928 he was eventually made the first Professor of Plant Pathology in Britain.[12] Brown was able to move the field forward because he devised a simple semi-quantitative method for assessing macerating activity. He assayed the activity of an extract by immersing in it discs of turgid turnip tissue (10mm diameter, 0.5mm deep); activity was taken as the reciprocal of the mean time taken for loss of coherence when that of the control was taken as unity. Brown stated, "coherence was lost ... when the discs as tested by hand offered no perceptible resistance to a pulling stress". The assay was obviously crude but it was effective. A series of papers was published, appropriately, in the *Annals of Botany*. By using cultures of *Botrytis cinerea* that were only 24h old, Brown was able to obtain extracts that were many times more active than those made by his predecessors. The active principle could not pass through the minute pores in the wall of dialysis tubing, showing it was a large molecule and not simply oxalic acid as some others had been claiming. Consonant with this conclusion, extracts could not diffuse through plant cuticle so they had no effect on intact leaves or flowers, although germination of spores in drops of water on those surfaces could be enhanced by small molecules, i.e. nutrients, diffusing out of the leaf. (Of course, spores themselves could give rise to cuticle-penetrating growth.)[13]

In his earlier papers, William Brown called the macerating principle 'cytase', in accordance with the name given by Horace Brown and Morris (1890) to the enzyme they found was responsible for breaking down the endosperm walls of germinating barley grains. However, Brown and Morris' cytase was subsequently found to be a hemicellulase (the first true pectinase was found, also in malted barley, by Bourquelot and Herissey in 1898) and in his later papers, William Brown adopted the name pectinase for his macerating principle. The first demonstration that enzymes are proteins was not made until 1926, when Sumner prepared crystalline urease from jackbean, and it was not until the end of the 1920s that the term 'enzyme' was widely adopted and there was reasonable consistency in the way different authors referred to particular enzymes and their substrates.

After 1928, Brown's only endowment to this field of research was his Ph D student, RKS Wood. In the 1950s, Wood was at the front of what was becoming a rapidly moving area whose importance was at last recognised. The diversity of pectic substances and the enzymes that degrade them was being appreciated (see Box 5:1). Brown had been unable to decide whether macerating activity and lethal activity were one and the same thing, or whether the first led to something, perhaps a toxin, responsible for the second. Now it became clear they were the same; by destroying the middle lamella, plus any plasmodesmatal strands connecting adjacent protoplasts, pectic enzymes rendered the plasma membrane (outer cell membrane) incapable of carrying out its normal processes of osmotic regulation, and the cell died in consequence. Within ten more years came the first evidence that oligosaccharides derived from the shortening of pectic polysaccharides chains in plant cell walls, particularly oligosaccharides that are 10–12 monosaccharide units long, function as endogenous elicitors. That is, they are signal molecules with the potential to be recognised by the plant and to switch-on its battery of active defence responses. Within twenty more years it became clear that important among those responses is the capacity of the plant to synthesise PGIPs (polygalacturonase inhibitor proteins), proteins that will neutralise its attacker's pectic enzymes. Today, pectic enzymes are finally recognised as being at the very heart of the dynamic interactions which make up the battle between plants and their potential pathogens.

To return to Harry, his publication on 'The Lily Disease' in 1888 is all the more remarkable because much of the work was done during a time of personal tragedy. The move south from industrial Manchester to leafy Englefield Green had been followed in 1887 by the birth of a second daughter, Dorothy. However, just when it seemed Harry and Selina were entering a period of great happiness, little Dorothy died on Tuesday, 27 March 1888. She was only six months old. The circumstances of the child's death are revealed in a black-edged letter Harry wrote to Dyer on 4 April,[14] "The child was accidentally smothered by her nurse, and you will understand how the inquest and funeral, and the necessity of attending my wife, deranged all my plans." Infant mortality was much more common in the 19th than in the 21st century although the epidemic diseases which killed most infants were much less common among the better fed, better housed middle classes, such as the Wards, than among the working classes. The event must have been all the worse for being unexpected; there was no hint that the child was ill (although today we might wonder whether she was a victim of Sudden Infant Death Syndrome (SIDS) or 'cot death'). The grief must have been hardly bearable, as only parents who have lost children will truly know.

His reaction to this single incident probably tells us more about Harry, the man, than any other in his life. On the surface, in his letters to Dyer, who was by now an old friend, emotions and feelings were kept to a minimum. His letter of 4 April, only eight days after Dorothy's death, starts and ends with botanical observations. On 11 April, he thanked Dyer for his sympathy, adding "... it has indeed been a terrible blow, but work and time will, we hope, enable us to forget the shock".[15] His letter immediately went on to discuss matters of the Council of the Linnean Society. It is not clear what 'work' he had in mind for Selina, but he threw himself into his investigations of the lily disease. Below the surface lay strong emotions that hard work could not drive away. His old friend, Bower, recalled how he had visited Ward at Cooper's Hill during the summer of 1888 and remonstrated with him for not taking his proper holiday.[16,17] In the following winter, bereavement and overwork caught up with him; Harry finally broke down. He had to stop, take stock and come to terms with his feelings.

He had had too to support Selina through her grief and, doubtless, it had become extremely difficult to sustain anything approaching normal family life. In the event it was Harry's mother-in-law who came to the rescue. Elizabeth Caroline Kingdon (nee Henderson) was a woman of strong opinions bluntly expressed, particularly about religion, and her relationship with her son-in-law had never been an easy one. Writing from Ceylon to his own "Darling mother", three years before marriage, Harry was blunt about his future mother-in-law. "I have a pretty fair estimate of that lady's worth, and know full well what a blunderer she is in many respects. That she is really kind and fond of her children I have had abundant opportunities of seeing – more than you have, remember – and of her wish to be and act as a good woman I am also convinced. Her great want is 'tact' – that peculiar weapon by which the lucky possessor is enabled to overcome more enemies than with sword or pen."[18] Like many dutiful daughters, Selina seems to have borne the awkwardnesses of her mother's character with forbearance, striving always to minimise any friction between her husband and his mother-in-law.

At the core of the problem were Elizabeth's strict religious beliefs. She was a member of the Plymouth or Christian Brethren. This Fundamentalist sect blossomed in the west of England, particularly in the city of Plymouth, after being carried there from Dublin in about 1830. (In its early years, up to 1,000 people attended the services, or 'Assemblies', at which there was neither a presiding minister nor a set form of devotions and which were sometimes held out of doors.) Regular daily prayer was at the center of Brethren practice, as was a weekly re-enactment of the Last Supper. Life at home at 'The Laurels' was in Winifred's words "... a good old mix-up. Heaven (up in the sky) and Hell (somewhere down below), angels, harps and wings from nurses; strict Sabbath-keeping and prayer-saying from Granna [grandmother], and a certain amount of the same from governesses; and at the back of it all, Father's rationalistic and scientific outlook, with a wise word in season regarding some of their superstitions; and Mother's tendency to shy away from the subject altogether."[01]

Winifred described her father Harry as being agnostic, rather than atheistic. Like his wife, he too had grown up in a family with strong Christian beliefs. His mother's occupation shortly before her marriage was described as 'seller of bible tracts' and his father's commitment to the church and to church music was part of the fabric of the family. It seems Harry probably developed a dislike for the more formal aspects of organised religion: there was no violent revolt against Christianity. Indeed, in 1894 he wrote a small textbook, *Diseases of Plants*, for the Society for the Promotion of Christian Knowledge (Fig. 5:4). Harry had inherited from his father a liberal, Whiggish, attitude to politics and society; as he grew older these views were increasingly permeated by the rationalistic outlook of a scientist.

To return to Harry's erstwhile difficult mother-in-law. Following the death of baby Dorothy, she proposed that Selina and the children should spend the summer of 1889 with her, away from the sadness at The Laurels. She rented a house at La Tresne, near Bordeaux, and there she, her daughter and grandchildren, not to mention 'Bean' the dog, enjoyed a long and relaxing holiday (Fig. 5:5). Relaxing that is except for the time when Frank was expelled from the local convent school for a number of misdemeanors. The final straw was when, in his own words, "I danced into the chapel during Mass, to fetch a little girl of whom I was very fond, to play hide-and-seek with me in the grounds." The Mother Superior told Selina, "Il a le diable dans ses jambes" (He has the devil in his legs). Selina was very ashamed of Frank for a least a week. His spirit of adventure was never quelled although, ultimately, it would be his making and a source of great pride for his mother. In the autumn, the family returned to Harry and 'The Laurels'. Selina's spirits were by then largely restored, but such tragedies are never completely forgotten. There were no more children.

SOCIETY FOR PROMOTING CHRISTIAN KNOWLEDGE. 3

THE ROMANCE OF SCIENCE.

Small Post 8vo, Cloth boards.

DISEASES OF PLANTS.

By Professor MARSHALL WARD. With Numerous Illustrations. 2*s.* 6*d.*

THE STORY OF A TINDER-BOX.

A course of Lectures delivered before a Juvenile Auditory at the London Institution. By CHARLES MEYMOTT TIDY, M.B., M.S., F.C.S. With Numerous Illustrations. 2*s.*

TIME AND TIDE.

A Romance of the Moon. By Sir ROBERT S. BALL, LL.D., Royal Astronomer of Ireland. With Illustrations. 2*s.* 6*d.*

SPECIFIC SUBJECTS.

Fcap. 8vo, 64 pp., Limp Cloth, price 4d. each.

ALGEBRA.

By W. H. H. HUDSON, M.A., Mathematical Lecturer, and late Fellow of St. John's College, Cambridge.

(ANSWERS TO THE EXAMPLES GIVEN IN THE ABOVE, *Limp Cloth*, 8*d.*)

EUCLID.

Books 1 and 2. Edited by W. H. H. HUDSON, M.A., late Fellow of St. John's College, Cambridge.

ELEMENTARY MECHANICS.

By W. GARNETT, M.A., Demonstrator of Experimental Physics in the University of Cambridge.

PHYSICAL GEOGRAPHY.

By the Rev. T. G. BONNEY, M.A., F.G.S., &c., Fellow and late Tutor of St. John's College, Cambridge.

Fig. 5:4. This small book written by Harry, *Diseases of Plants* (1894), price two shillings and sixpence, was one of a series entitled *The Romance of Science* published by the Society for Promoting Christian Knowledge. The 193 pages of text were richly illustrated, often with original drawings by Harry.

Fig. 5:5. Winifred and Frank on holiday at the Villa Beauséjour, La Tresne, near Bordeaux, probably in 1889, although Winifred's inscription (made many years later and probably in error) is "April 1891". (By permission, the Royal Botanic Gardens, Kew.)

Harry returned to his self-imposed punishing schedule of work. Life moved on and, gradually, happiness returned. Harry was a great letter writer and since his days in Ceylon he had built a wide circle of professional contacts through his correspondence. His initial aim was clearly to advance his career and to cement a place for himself in the botanical 'Establishment'. However, true friendships sometimes sprang from these connections and the Poultons in Oxford became family friends in just such a way. Edward Poulton was an entomologist, who was made the Hope Professor of Zoology in 1893, and a man with whom Harry often corresponded about their mutual interests in the pests and diseases of plants. Each family was entertained at the other's home and joint visits to events in London were organized. Winifred remembered holidays in Wales were sometimes taken with friends.

Harry was becoming involved in new areas of research and these too sometimes led to family friendships. Thus, the Wards became good friends of the Horace Browns: the connection was that Horace was a biochemist at Worthington's of Burton-on-Trent, the center of British brewing industry, and brewing was in industry with which Harry was getting involved.

After the major life events of the 1880s, Harry seems to have been in good health and spirits by the early 1890s. In one letter,[19] dated 1 October 1991, he requests that Poulton should, "Please tell Miss Shepherd my hands are still tingling for – or perhaps from – a game at fives". (Fives is a game not unlike squash, but in which the ball is struck with gloved hands, rather than a racket, against the walls of a three-sided court.) Typically, even ominously, he adds, "I am in the thick of work again: would much rather be careering along the Sandown road on a tricycle with you all." Throughout the 1880s, and even into the early 1890s, he was at his most productive and original, in spite of the major life events. He maintained a healthy, if precarious, balance between his professional and his personal life. After this period, as we will see in the next two chapters, he took on ever more demanding tasks and responsibilities. More and more those commitments took him away from his home and his family and, all too soon, they helped to kill him.

Chapter six: Ginger beer and a loss of focus?

"... it is a particular belief of mine that many of the questions raised in the special study of fermentation by you practical brewers are the same as raised by us – shall I say theoretical botanists?"

From *Some Brewing Botanical Problems*:[01] an address by Harry Marshall Ward, D Sc, FRS, FLS, Professor of Botany in the University of Cambridge, to members of the Federated Institute of Brewers after dinner at the Criterion Restaurant, London, on the night of Thursday, 12 May 1898.

Harry was at the pinnacle of his career in 1898. He was a Fellow of the Royal Society and just three years earlier he had left Cooper's Hill to take on the Chair of Botany at Cambridge. He was able to take a senatorial view over the progress of botany and microbiology, and to look back with some satisfaction at his part in that progress, for he had made an enormous contribution to both subjects. The years from 1885 to 1895, spent in the stable environment of Cooper's Hill, had been Harry's most productive in terms of both the quality and quantity of the original research he completed. He had matured as a scientist and his interests had broadened. He had recognised that industry and science could profitably learn from each other. This was nowhere more true than in the links between the brewing industry and the emerging sciences of microbiology and biochemistry.

Based on his first hand practical experience of the microbiology and biochemistry of brewing, he had by 1898 written several papers for journals linked to the brewing industry, journals whose readers included a high proportion of practical brewers. Not least among his papers was a brilliant series of short educational articles entitled 'The Fungi of Fermenting Vessels, Vats, &c.' written for *The Brewers' Journal* (1890–91).[02] He had also corresponded regularly with many of his fellow diners at The Criterion and, as seen in the previous chapter, was a particular friend of Horace Brown, one of their leaders. And those were the reasons, apart from the offer of a fee (which was always welcome) and an excellent meal at one of London's top restaurants (Fig. 6:1), why Harry was the honored guest of the brewers.

Beers had been brewed throughout the world for centuries but the (micro)biology of beer production and spoilage were still in their infancy (see Box 6:1) when Harry's interest was first kindled in 1887. It was not commercial brewing that first attracted his attention and, as with so much else in his life, the impetus stemmed from Thiselton-Dyer. One day in 1887

Fig. 6:1. Piccadilly Circus, London, pictured soon after 1900. The Criterion Restaurant is seen to the right of the Circus.

when visiting his old mentor, who two years before had succeeded Joseph Hooker as Director of the RBG at Kew, Harry's eye fell on bottles of ginger beer in the great man's study. "There is a thing you have to worry out", said Dyer.[03] Whether out of gratitude for past favors, or in anticipation of future favors, or simply because he was in such awe of Dyer that he couldn't say 'No', Harry agreed to investigate the nature of the 'Ginger Beer Plant'.

It seems that in the 1880s the Ginger Beer Plant had become a popular subject of speculation among biologists. A 'Plant' was exhibited in 1883 at a meeting of the Essex Field Club, a club attended by leading mycologists of the day, such as Mordechai Cubitt Cooke, who were all perplexed and confused as to its true nature.[04] Also in 1887, before Harry's critical visit to Dyer, Bayley Balfour, who at the time was Professor of Botany in Oxford, had exhibited specimens of the 'Plant' at a meeting of the Linnean Society. Balfour had likened Ginger Beer to 'The Kephir of the Caucasus', a traditional drink of the Caucasus prepared by the actions of a fungus, known as 'kephir grains', on cow's milk: when kept in closed vessels the milk fermented, producing a thickish white, undoubtedly alcoholic, liquid – the 'Kephir'. Some said Ginger Beer was brought to Britain by soldiers returning from the Crimean War. Whatever its origins, it was a popular, highly alcoholic, summer beverage. However, since it was produced either in very small quantities domestically, or in only modest amounts industrially, its improvement had little commercial potential. Harry's starting point was a liquid suspension that was neither uniform – closely guarded recipes for ginger beer having been passed down within families from generation to generation – nor produced under sterile conditions. It was nothing short of a microbiologist's nightmare. Even Harry was to admit, "… had I known how

Box 6:1. Yeasts, alcohol and beers

Yeasts are fungi but, unusually, they are unicellular rather than filamentous. They occur naturally on the surfaces of plant leaves, stems and, particularly, on fruits and seeds, where they live on the meager supply of nutrients leaking from the plant. When a plant dies, or its fruits are harvested and mashed, these yeasts proliferate and there is a concomitant increase in the rate at which they metabolise the sugars contained in those plant organs to ethyl alcohol and carbon dioxide. The chemistry of the process, which is known overall as 'fermentation', was the starting point for the career of Louis Pasteur.

$$\underset{\text{Glucose}}{C_6H_{12}O_6} \Longrightarrow \underset{\text{Ethyl alcohol}}{C_2H_5OH} + \underset{\text{Carbon dioxide}}{CO_2}$$

The 'yeast' in traditional beers was not a single pure strain of the species *Saccharomyces cerevisiae*, but was composed of many different strains, each growing at a different rate and secreting different flavour compounds (up to 400) into the beer. In Britain, brewers learned to achieve consistency between batches by inoculating each new batch with the froth skimmed from the top of an old batch when they judged it contained just the right mix of strains. Beers in Continental Europe evolved along a different route as brewers adopted a Bavarian method that used a 'yeast' which during fermentation sank to the bottom, rather than rose to the top, of the vat. Thus, whereas 'ale' produced by a top-fermenting yeast was the beer of Britain, 'lager' produced by a bottom-fermenting yeast (at a lower temperature) was the beer of Europe and, subsequently, of North America. Arguably, the greatest contribution that microbiology made to brewing occurred in Copenhagen in 1888 when Emile Hansen isolated from the sludge at the bottom of such a lager beer a pure strain of a yeast, which was called *Saccharomyces carlsbergenesis* to distinguish it from *S. cerevisiae.* With such pure cultures came reliable products. *S. carlsbergensis* was exported worldwide, earning vast revenues for the Carlsberg Institute in which Hansen worked.

long and difficult a task I had set myself, the attempt would possibly have been abandoned at an early stage".[05]

Some five years and over 2,000 cultures later, and having avoided death or disfigurement from shards of glass flying from vessels exploding under the mounting pressure from CO_2 generated within them by fermenting cultures, Harry produced his 78-page report. He described how his method was to isolate the various micro-organisms and then, by putting them together again in pairs, to test their ability to re-create a ginger beer plant. He had found there were two essential microbes in the semi-translucent tapioca-like lumps common to all Ginger Beer ferments, a yeast, *Saccharomyces pyriforme*, and a bacterium, *Bacillus vermiforme* (now called *Brevibacterium vermiforme*).

Remarkably, he was able also to disentangle evidence to show that there were two other "constant but not essential forms found in specimens examined, … *Mycoderma cerevisiae* (now *Candida cerevisiae*), the well known agent of the "mould" on sour beer … while the other is the vinegar bacterium, *Bacterium aceti* (= either *Acetobacter* or *Gluconobacter*)". In addition, "Foreign intruders included species of *Bacillus, Micrococcus, Saccharomyces, Torula, Dematium* and *Penicillium*". We know now that whitish-translucent clumps, comprised of comparable cocktails of symbiotic micro-organisms, occur in other domestically brewed beverages from around the world; the clumps are variously called 'Wasserkefir', 'Japanese crystals' or 'graines vivantes'.

Most importantly, Harry recognised that the ginger beer plant had some of the characteristics of the naturally symbiotic gelatinous lichens with which he was already familiar from his botanising in Ceylon (chapter 1). Just as in lichens, where one partner (the fungus) provided a framework within which the other (the alga) lived, in the ginger beer plant the bacterium secreted a mucilaginous sheath which formed a matrix supporting the yeast cells. In the artificial, just as in the natural symbiosis, physiological activities of the two microbes were complementary. Harry stated merely, "the Schizomycete [bacterium] is favoured by obtaining some substance or substances directly … [as] … they leave the yeast cells", while the yeast "benefits by these substances being removed and destroyed". His successors were bolder in clearly identifying the substance in question as carbon dioxide. Presumably, the bacterium utilised carbonic acid (the soluble form of carbon dioxide) and from it synthesised other organic acids, and possibly vitamins, essential for the yeast.

In a way, and almost certainly in a way that he had not planned, Harry did manage to reap some reward for his investment of time in Ginger Beer. Academically, he emerged with a refined concept of symbiosis and, among other articles for the *Transactions of the Brewing Industry*, he was able to write authoritatively on 'Symbiosis and Symbiotic Fermentations'. He had also, of necessity, become an authority on the taxonomy of the Schizomycetes. (Schizomycete was a term used to describe one celled organisms; it included bacteria and yeasts). He was made an honorary member of the Institute of Brewing which then, as now, was an honor reserved for the most distinguished scientists connected with the industry, for example, Louis Pasteur. More generally, the celebrity he gained must have been heartening for a man who sometimes lacked confidence in himself rather than in his science. The small income from consultancies, which came his way from the brewers, helped to pay for the children's governesses and school fees (chapter 5). Even Selina was drawn in, receiving from "a lady in Paris a number of grains of a body that looked like boiled sago", accompanied no doubt by a request for her husband's help.[06] The Parisienne had, in her turn, been given the grains by a missionary from Madagascar. It is not recorded whether the missionary wanted to encourage or discourage the consumption of the products of the mysterious grains.

In his Criterion address, we can sense Harry's excitement at the prospect that microbiologists and brewers could create their own symbiosis. Carefully using his knowledge of the latest advances in mycology to select examples that were relevant to the working lives of his audience, he told them how it was "recently shown that 'corked wine' owes its peculiar flavor to the inroads of *Penicillium,* and also that *Mycoderma* (*Candida cerevisiae*) burns off the alcohol and induces flatness"; he added that 'ropy beer' arises because of the mucilaginous slimes extruded into it by a number of spoilage micro-organisms. He then went on to speculate that other recent advances in mycology might one day have direct applications. He described, for example, how many filamentous fungi can, under the right environmental conditions, be induced to grow in a yeast-like form, a phenomenon now known as yeast-mould dimorphism. (Such conversion from a filamentous to a more mobile yeast-like state explains the rapid spread in the vascular system of their host of fungi as diverse as the human pathogen, *Histoplasma capsulatum*, and the Dutch Elm Disease fungus, *Ophiostoma ulmi*.) He suggested the opposite change in morphology was also possible, that the true yeasts have the potential, and one day might be persuaded, to grow in a normal filamentous form. His proposal was correct and, although the phenomenon has yet to be commercially exploited in brewing, knowledge of the factors triggering the change from yeast to mycelium does help us understand, for example, how *Candida* infections growing in yeast form on human mucosal membranes can under high concentrations of sugars in their environment, as in diabetic hosts, change to a mycelial form which invades deep tissues.

Just as at the beginning of the 21st century *S. cerevisiae* has become a tool favored by molecular biologists, being, along with man, one of the first organisms to have its complete genome sequenced, so, at the dawn of the 20th century, it was a tool favored by those unlocking the secrets of enzymology. In the audience at The Criterion was Harry's good friend, Horace T Brown, who, with his collaborator Morris, had done so much to explain the malting process by demonstrating how germinating cereal grains secrete [the enzyme] diastase to convert their store of starch into glucose, the energy and carbon source needed by *Saccharomyces* for growth and fermentation. Harry next described at length the revolutionary discoveries of Buchner which, in terms of unravelling metabolic pathways, took Brown's work further. Pasteur had proposed that fermentation took place only inside living cells, but Buchner (1897) overturned this idea when he showed that it would proceed in sap after it was extracted under high pressure from *Saccharomyces*. He called the active principle of this process 'zymase'. The discovery was puzzling, admitted Harry. (We must remember that, as explained in chapter 5, contemporary understanding of the true nature of enzymes was still very confused. Moreover, the few enzymes isolated to date had been soluble and either secreted, before functioning outside the cell, or easily extracted from the cell that produced them.) Zymase was neither readily soluble nor secreted; it could be extracted only with difficulty, leading Harry to speculate that it might be of protoplasmic nature – the idea of 'bound enzymes', attached to structures such as cell walls, was still some way off into the future.

The special relevance of Buchner's work to the brewers was that zymase was active outside the cell and that its extracellular activity could be controlled, for example, by the addition of sugars, thus raising hopes that brewing might eventually divorce itself from the demands and sensitivities of living yeasts.

To return to Harry and his productive decade at Cooper's Hill, the other major outcome of his work on the Ginger Beer Plant was an invitation issued by the Royal Society in 1892 to join Percy Frankland in producing a report for its Water Research Committee. The project was to compare the quality of Thames river water with that from other natural sources, regarded as relatively clean, such as Loch Katrine in The Trossachs of Scotland. Again, he could not have foreseen the full volume of the work in which he was to become entrapped. He was certainly not adequately compensated by the honorarium of £100 he received in each of the three years he was involved with the project.

In Europe and in North America, the nineteenth century was characterised by the mass movement of people from the countryside to the growing cities. As we have already seen in chapter 4, this aspect of the Industrial Revolution was seen nowhere better than in Great Britain with huge increases in the population of manufacturing bases such as Manchester, Nottingham and London. Sources of potable water, and systems for the disposal of sewage, were stretched to their limits – and beyond. The classic water-borne diseases, cholera and typhoid, were common; in Britain there were major epidemics of cholera in 1831, 1848 (62,000 lives were lost in the latter outbreak), 1853 and 1866, and of typhoid in 1837, 1847, 1855 and 1868. In 1886, eels, living and dead, emerged from taps in some parts of London. 'Improvements' in the homes of the middle and upper classes, such as piped water and flushing toilets, merely served to exacerbate the problems, since they increased discharges to streams and soak-aways.

At the heart of the problem lay the fact that water was supplied by private, profit-making companies – eight in the case of London. In the absence of sufficient underground, naturally filtered deep-well sources, these companies drew water from, and discharged sewage back into, the tributaries of the Thames that wound their way through London. The activities of these companies were both uncoordinated and self-serving. Until parliamentary pressure began to grow in the 1880s for them to be taken over by a public authority, when the companies saw the good sense in making common argument, they applied widely different standards of analysis and purity.

The analysis of potable waters had been a concern that had been almost private to the Frankland family since 1865 when Edward Frankland, having left Owens College, succeeded AW Hoffmann at the Royal College of Chemistry, not only as examiner for chemistry at the Department of Science and Art but also as Official Analyst of the London water supply. Frankland was heir to the long tradition that water analysis was the business of chemists. The chemists were a powerful group for, as Colin Russell shows in his excellent *Edward Frankland: Chemistry, Controversy and Conspiracy in Victorian England*,[07] science in mid-Victorian Britain was virtually equatable with chemistry. Moreover, there were sound economic reasons for the *chemical* analysis of water, not least the need to determine its 'hardness', because a high concentration of calcium and magnesium salts was wasteful of soap, and their soluble bicarbonates often led to dangerous blockages of steam pipes by producing deposits of insoluble carbonates. Thus, the chemical characteristics of each water could be important tools for the water companies in their marketing. Edward Frankland analysed waters not only from the metropolis but also from, amongst other places, the royal palaces of Windsor, Sandringham and Balmoral – all on a strictly commercial basis.

Although John Snow's classic work following the 1854 outbreak of cholera in London had shown that the disease was transmitted by water specifically polluted by the excreta of a cholera victim, he did not commit himself to the unequivocal view that a living organism was the sole cause of the disease. Instead, his ideas of causation were strongly influenced by the prevailing 'zymotic theory', owing much to Liebig, the German chemist. This proposed that decaying organic matter caused disease by passing the decay processes to living tissues. The first isolation of bacteria, and the growth of a science around them, did not instantly change views.

By the mid-1880s, however, most water analysts in Britain had accepted that bacteria, or other living entities, caused water-borne diseases, but they had little idea how to translate this knowledge into any practical measures to improve drinking water. It was the fashion to isolate bacteria from water sources and to count colonies cultured on nutrient media in glass dishes, but there was little agreement as to what the numbers meant. Much practical bacteriology was done by untrained or unskilled practitioners. There was little realisation that each different growth medium was partially selective, favoring to some extent its own peculiar assemblage of micro-organisms. Fully selective, species-specific media, which in contrast to partially selective media are very useful, had not yet been developed. Consequently, there were few satisfactory means of distinguishing the multifarious species of bacteria from each other, let alone of separating the pathogens from the non-pathogens. The new science of bacteriology was, for the moment, posing many new questions but answering few. Indeed, it was still possible for some eminent men to defend the view that the microbes found in polluted waters were merely feeding on the rich organic food-base to be found there, i.e. the microbes were saprotrophs not parasites. The bacteria causing cholera and typhoid had still not been isolated, allowing some serious analysts still to doubt their existence. Edward Frankland, protecting the view that the chemist could provide all the answers required, but tacitly acknowledging that the causes of cholera and typhoid were spread by sewage-contaminated waters, modified his standard method of analysis to incorporate measurements of the nitrogen in water – nitrogen levels being in his view indicative of 'previous sewage contamination'. The arguments and, even more so, his methods of analysis were extremely controversial, but Frankland's business flourished. As often happens when a business grows, work was farmed out, and Edward's son, Percy, was brought in to help.

Percy had been educated at the Royal School of Mines (1875) where his brilliance earned him a scholarship to study medicine at St Bartholomew's Hospital. However, his father wanted him to be a chemist. To re-inforce his wishes, Edward gave to Percy for his next birthday a copy of his book, *Experimental Researches,* and a cheque for £130, the value of a scholarship. Percy knew he had lost this battle and within days he was in Würtzburg studying for a Ph D in organic chemistry. On his return to Britain, he worked for a while as his father's assistant at Grove House, the new laboratory to which the Official Analyst's work moved in 1882. Percy was every bit as determined and strong-willed as his father. They clashed over both scientific and family matters and, almost inevitably, in 1888 Percy left Grove House. At the time he began his collaboration with Harry, he was Professor of Chemistry at Dundee but in 1894 he moved to a similar post in Birmingham.

Percy was not appointed to the Royal Society's Water Research Committee as a chemist, however, but rather as a bacteriologist. He had embraced the new science in more ways than one for he had married Grace Toynbee FLS, an experienced bacteriologist and therefore an exceptional woman for her times. Percy clearly learned much from Grace, and she provided inestimable practical support to his studies, but she was not his only source of expertise. He probably learned Koch's plate culturing techniques (see Box 6:2) through a working laboratory set up in London in 1884 as part of the International Health Exhibition though, it is believed,[08] he also trained directly under Robert Koch some time in the 1880s.

After separating from his father, Percy continued to earn a considerable income through work for the private water companies. Percy's motivation was to show to the public, and the government, that polluted waters could be purified by passage through filters such as sand beds. Quite naturally, he was the expert they chose to provide them with the evidence, so there was great potential for biased reporting.

Percy was well aware, however, of the need to be seen by the scientific community as the impartial expert. To this end he was a master of equivocation. In his 1887 address to the Society for Chemical Industry he said his tests indicated "the efficiency of filtration" and "variations in the amount of organic life present". He went on, "No conclusions whatever, as to the relative excellence of various waters, are to be drawn from their greater or lesser freedom from micro-organisms, any more than it is possible, on the strength of chemical composition, to say that one water is more wholesome than another. ... On the other hand, these determinations undoubtedly do indicate what would be the probable fate of any harmful organisms gaining access to the sources of supply, and what is their relative chance of reaching the consumers; for that method of treatment which abolishes the largest proportion of organisms of all kinds is also most likely

Box 6:2. Robert Koch: a proud Prussian pioneer of microbiology

(a) pioneer of solid, rather than liquid, media for the isolation of microbial colonies or pure strains

Koch's new method was announced at the International Medical Congress in London, August 1881,[09] but according to Frankland and Ward (1892),[10] "... the method was merely an improved adaptation of one introduced long before [1874] by the botanists Brefeld and Klebs". Frankland and Ward added, however, "To Koch belongs the credit of having applied these plates to the purpose of separating the various colonies of micro-organisms." At first, gelatine was poured onto sterile plates covered by glass domes, but one of Koch's assistants, Richard Julius Petri, greatly improved the technique. He developed the closed glass 'Petri dishes' used ever since by all microbiologists. Remarkably, Percy Frankland published in 1886[11] a description of what we would call a Petri Dish; this was one year before Petri published his account of the dish!

Koch's first growth media were based on gelatine but some bacteria were found to liquefy it and on hot days it melted spontaneously. He improved the technique by replacing gelatine with agar, a seaweed-based product that allows a cooling liquid medium to gel at 45°C but prevents it re-melting below 90°C. Colonies of microbes could now be grown on top of the media and, thus, more easily isolated and manipulated.

(b) the spread of anthrax caused by *Bacillus anthracis*: a topic as newsworthy now as then

Koch demonstrated that anthrax could be produced by the transfer of infected material from one farm animal to another. Pasteur confirmed 'the germ theory of disease' by using membrane filtration to prove that it was the bacilli that transferred the disease, not the medium in which the bacteria grew. Koch also demonstrated the ability of the bacillus to produce heat-resistant spores, which had the ability to germinate and infect animals: the soil, not water, is a natural reservoir of infection. He thus explained what had long been a mystery; why healthy animals continued to become infected long after diseased animals had died and been removed from the pasture.

to abolish any pathogenic forms should they be present." The truth, as he was to admit later in the 1890s, was that the water filters, which usually consisted of only a few feet of sand, were only effective if they were maintained to the highest standards.

Little was known about the abilities of water-borne pathogens in general, and the bacteria that form resting endospores in particular, such as *Bacillus anthracis*, to survive in the natural environment, but the subject was beginning to be addressed by the late 1880s in a number of German and other continental laboratories. In Britain, Percy was getting involved: he investigated the survival rates of six organisms, including the *comma vibrio* (thought to be the cause of cholera), *B. anthracis*, and the putative typhoid organism. Despite the highly standardised and artificial conditions he employed, the survival period of the microbes differed significantly even in the laboratory. It was still a complete mystery how they would survive in the continuously variable natural environment. At this stage the Royal Society commissioned its report.

The Royal Society's case for involving Percy Frankland was obvious, the value of his expertise outweighing any objections that might have been raised concerning his financial links with the water industry. But why was Harry Marshall Ward chosen? Probably because of his recently acquired expertise in the taxonomy of Schizomycetes and because of the inventiveness and thoroughness that characterised his laboratory work. Harry's experience of the complexities of mixed microbial communities, as in beers, meant that he was exceptionally well qualified to investigate the microbiology of what were now being appreciated as mixed communities in water.

It is not clear where Harry's expertise in bacteriology stemmed from, or where his interest in waterborne bacteria originated. One possible source of inspiration for his interest was Robert Angus Smith, a talented but maverick scientist who did not fit comfortably into the scientific establishment. Back in 1882, soon after Harry's return to Owens College (chapter 4), Smith presented to the Manchester Literary and Philosophical Society an account of novel experiments using gelatine to culture the bacteria in water. Harry had an extraordinarily retentive memory and he would not have forgotten Smith, a man who was unusual in being fully conversant with the views of Koch and Pasteur and a man who saw the implications of their work for water analysis.

Collaboration, teamwork and multi-author publications have become the norm for today's biologists, in spite of the inter-personal stresses and strains that are involved. In contrast, biologists of the late nineteenth century typically worked alone and published alone. Collaboration was therefore both novel and challenging for Harry. He was going to have to work with the son of one of the most famous scientists of the age and by all accounts a man who knew his own mind. Percy's bearing was formal and unsmiling (Fig. 6:2) and he had a reputation for being strong-willed, but he was also described as being impetuous, generous and warm-hearted, ever eager to combat injustice.[12] He seems to have engendered strong feelings in others, sometimes negative, sometimes positive. At the outset of their working relationship the omens were good because Harry and Percy shared several experiences and interests. For example, they had both been taught at South Kensington, although not at exactly the same time since Harry was four years older than his collaborator. Each passionately believed that awareness of the history of their subject was a prerequisite for its proper understanding. They also shared interests away from science, not least their love of the wilder British countryside of the coast and mountains. Percy was a passionate and renowned cyclist. In his student days in Würzburg he had amazed his German

Fig. 6:2. Professor Percy Frankland CBE, FRS. (By permission, the Royal Society).

friends by riding a 'penny-farthing' bicycle. As we know already (chapter 5), Harry too greatly enjoyed cycling. Another shared interest was music. Percy, like Harry, was keenly interested in music – his favorite composers being Mozart, Wagner and Tchaikovsky – but, unlike Harry, the only 'instrument' he could play proficiently was the pianola (not unlike a modern barrel-organ); self-deprecatingly, he liked to joke about the 'expression!' he got out of this purely mechanical device.[Juliet Frankland pers comm]

We do not know how well their collaboration developed on a personal level but there is a suspicion that it could have been closer. For example, the published writings of the two men were, in the main, not well integrated, and Harry clearly did more than half the work although the men received the same honoraria from the Royal Society. Also, it seems significant that in the book, *Micro-Organisms in Water*, which Percy and Grace Frankland published in 1894,[14] the same year as the final report to the Royal Society,

Harry's work receives scant mention or credit in those chapters in which it might be expected to feature.

The outcome of Harry and Percy's collaboration was a series of five reports published sequentially in the *Proceedings of the Royal Society*, sometimes with the contributions of the two authors being separated into different parts. The series started, as might be expected, with what was essentially a review of the literature. This First Report[10] is a mere 97 pages long, jointly written, and was published in 1892. Frankland and Ward concluded that a clear picture emerged from what was already known; they wrote that it can be agreed "in all ordinary waters the rule is that the pathogenic forms [or microbes] die out sooner or later". And then, looking forward to the practical part of the project, they set themselves specifically the task of defining the effects of nutrients, temperature and light upon the survival and multiplication of these organisms, both spore-formers and non-formers. They explicitly recognised an important truth, that competition between species in mixed microbial communities of pathogens and saprotrophs, particularly in the organic muds into which suspended microbes may eventually settle, could eliminate pathogenic species. In the next three years there followed three more reports, which described the outcome of their studies of natural waters, and included experiments using some of those natural waters and also variously sterilised 'pure' waters. They experimented with single species and, following the methods of Harry's work with ginger beer, they examined artificially created mixtures. Overall, they picked up the questions raised in the First Report but, oddly for the modern observer, there was no 'Executive Summary'; they made no attempt to reach either a final synthesis or to present an overview. Standard practices were not then as they are now, but maybe the two men were never sufficiently close to each other to achieve that meeting of minds which a joint final summary would require.

The Second Report[14] dealt with experiments concerning the vitality and virulence of *Bacillus anthracis* and its spores in potable waters. The choice of the organism was odd because *B. anthracis* is not a waterborne pathogen (see Box 6:2). It owed its selection to being "one of the hardiest and most refractory of living organisms, at least of the pathogenic kind". There was a brief joint introduction and conclusion, otherwise Frankland and Ward wrote separately, in Harry's case for 135 pages. The Third Report[15], also, had two sections; in his 80-page section, Harry took up the question of the action of sunlight on *B. anthracis* and other bacteria in Thames water. In the Fourth and final report[16] of 204 pages, there is apparently nothing from Frankland. In total, Harry wrote much the larger part of the reports. Considering the first four, he was responsible for a staggering 504 pages of text and figures. Harry's writing was never succinct, and although the standards of the day allowed more verbosity than is acceptable today, the length of the reports must have deterred many readers. Indeed, we will see (chapter 8) that it was noted disapprovingly by a number of his peers. In his defence, the 504 pages lay out every experiment and observation in the most minute detail. They guide the reader, by deductive reasoning, to conclusions which mark out Harry as a scientist who would be outstanding in any age. They are testament too to his energy and dedication. From the laboratory records it may be seen that the continuous cultures of micro-organisms which he set up had to be examined at all times of day and night. Nevertheless, except for some help from GE Cartwright Wood with the second report and from a Miss Hayward (University College, London) with the third, the experimental work, measurements and detailed drawings were his own. He even found time to collect Thames river water for Frankland.

The experimental work can be criticised. Most questionable is the applicability to water-borne bacteria of conclusions drawn from a naturally soil-borne pathogen. Another weakness is the lack of any formal statistical analysis, as revealed in phrases like the following, from the third report: "without attempting to lay too much stress on actual numbers [of bacteria], it is pretty evident that if we take the totals ... or means of three samples, we get at least some information as to ...". However, this last shortcoming was perfectly normal for the day and it did not diminish the importance that readers and the Royal Society attached to the reports. Although it might be an exaggeration to say that they laid the foundations of environmental microbiology, they certainly gave the subject great impetus and established a perspective that could not be ignored by those who followed. Embedded in the voluminous reports were, at least, two brilliant pieces of experimental work by Harry.

The first of these enabled him to confirm by experiment what Percy's father, Sir Edward, had argued with his adversaries so bitterly, so many years before – that biologically polluted waters can become sterile and safe to drink. The bacteriocidal properties of sunlight had been proposed some fifteen years earlier (Box 6:3) and, among those who accepted the idea, it was generally held that light at the blue end of the spectrum was most effective. However, it remained for Harry to devise experiments which demonstrated for the first time that it is the ultra-violet wavelengths of sunlight which kill significant numbers of microbes, even spore-formers like *B. anthracis,* the microbe at the centre of Harry's studies.

The way Harry planned his experiments was exemplary. He based them on his exceptional knowledge of what other researchers in the same area had recently reported, but he determined to improve on their methods. He had little appreciation of dose-response relationships (Box 6:3) and made no attempt to quantify the fluxes of light energy with which he irradiated micro-organisms, but he took the best practical advice available from physicists in the early 1890s. He consulted two of the most eminent men, Sir Gabriel Stokes and Sir Oliver Lodge, who gave practical advice on the choice of glass filters which would transmit only particular wavebands of light – for example, Harry's filter 'Green 8' cut out most of the orange-red and blue-violet bands but transmitted green wavelengths – and, also, on the making of photographic records of bacteria exposed to different parts of the spectrum (Fig. 6:4 a, b). The final plan was Harry's and, again, it was he who virtually lived in the laboratory making precise measurements hour by hour. The

Box 6:3. Sensitivity to sunlight

Sunlight, or solar radiation, is a form of energy. Organisms differ in their sensitivity to the same amount of solar energy, i.e. each species has its own dose-response relationship.

Sunlight has the properties both of arriving in waves and of being delivered in packets, or quanta. It has a spectrum of wavelengths, those from 380 to 750 nanometers being detected by the human eye as different colours (Fig. 6:3). The eye cannot however see wavelengths from 310 to 380nm, comprising the bands called ultra-violet light. The energy carried by each quantum is inversely proportional to wavelength, meaning therefore that ultra-violet light is, relatively, highly energetic. Human skin is a good indicator of the energy content of the different wavelengths comprising sunlight. Thus, skin ageing is associated with the UV-A band (325–380nm), while sunburn is associated with the more energy-rich UV-B band (290–325nm). The UV-C band (< 290nm) in solar radiation doesn't reach the earth's surface because, fortunately, it is filtered out by the ozone layer in the earth's outer atmosphere but, when generated by lamps in the laboratory, it rapidly kills cells – and it is by generating UV-C radiation that germicidal lamps are able to sterilise anything from a laboratory bench to packaged foodstuffs. The biological effects of UV-radiation are dose-dependent, where:

$$\text{dose} = \text{flux (energy per unit time)} \times \text{time}$$

thus, a short period of exposure to short-wavelength radiation can be more damaging than long exposure to long wavelengths.

Note: the bacteriocidal properties of sunlight were first proposed by Downes & Blunt[17] in 1877.

conclusions were based almost exclusively on *B. anthracis*, so there was no direct attempt to resolve the question that different species might have different sensitivities, that is, they might have different dose-response relationships.[18,19]

Perhaps to satisfy his own curiosity, rather than any plan agreed with Percy or the Royal Society, Harry included some fungi in his experiments. Finding the colorless spores of some species, e.g. *Oidium* (a powdery mildew), were much more easily damaged than the pigmented spores of species such as *Botrytis* or *Nectria*, he correctly speculated that many of the orange and red pigments of fungal spores and fruiting bodies have been selected to protect those bodies from solar radiation. Unfortunately, he did not realise that different growth stages of the same organism, for example spores and mycelia, might have different sensitivities. But he did have the foresight to suggest that light from electric arcs might be used for the disinfection of hospitals.

The second piece of exceptional science was hidden in the very long, fourth report. (This was effectively the final report since the fifth report,

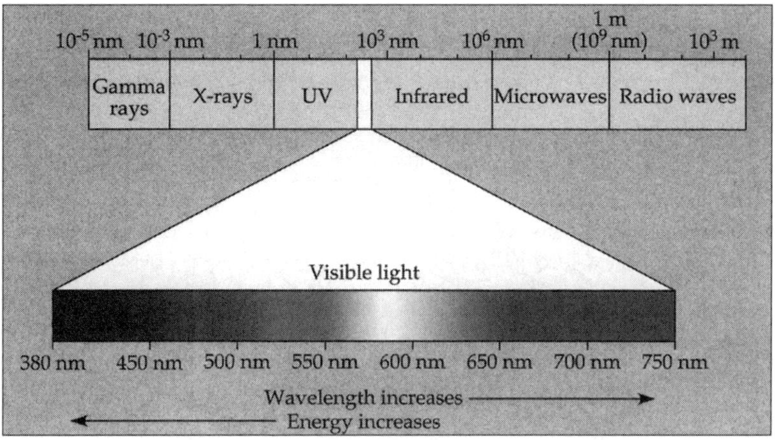

Fig. 6:3. The spectrum of sunlight. UV = ultra violet light is not detected by the human eye. (From NA Campbell, JB Reece and LG Mitchell, 1999, Biology. Benjamin/Cummings. Reprinted by permission of Pearson Education, Inc.)

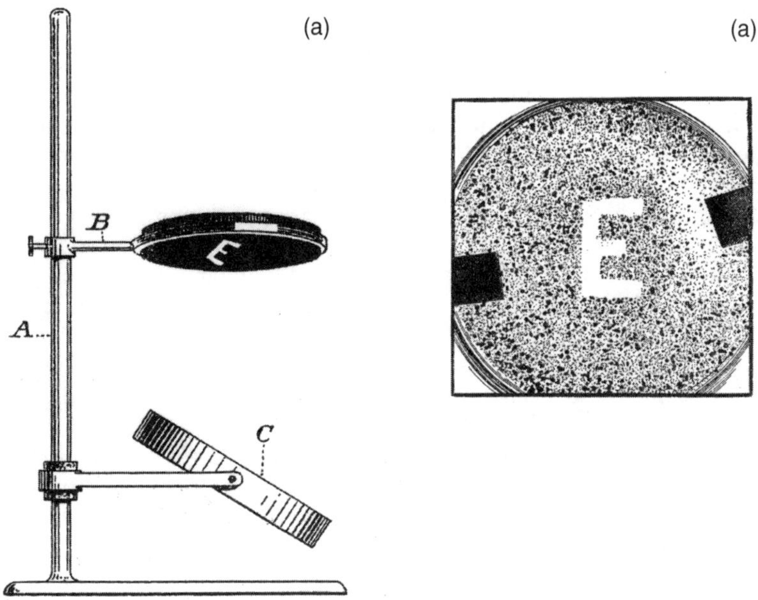

Fig. 6:4. (a) A Petri dish culture of the micro-organism, was held in a ring (**B**) on a retort stand (**A**). Sunlight was reflected by a mirror (**C**) onto the dish wrapped in black paper. Following an idea of Buchner (1892), a letter '**E**' was cut from the paper. Filters to select light of different wavebands were fixed in front of the '**E**'. Note: the glass of the Petri dish would have depleted the UV-component of any light reaching the micro-organism. (b) typical appearance of a plate, with no colonies in the '**E**', after exposure to light inhibiting growth of the micro-organism. (From Proceedings of the Royal Society of London 52, 393–400 (1892).)

which was written by Harry, amounted merely to a technical appendix describing in detail materials and sampling methods.) From the most detailed studies of the growth of *Bacillus ramosus* emerged a novel idea, which soon became accepted as a fundamental concept of microbiology, and which has often been called 'The Law of Doubling'. In its initial and simplest form it described growth in the size of populations of microbes which reproduce by fission, where the total number of cells doubles with each new generation. It is one form of the 'compound interest or exponential growth law' familiar to all who put money into a savings account; money invested grows continuously according to the time for which it is invested and the rate of interest.

$$K_t = K_0 \cdot e^{r.t}$$

Where K_t = capital, or population, at chosen time, t

K_0 = capital, or population, at time zero
e = 2.7 (correct to 1 decimal place): the base of natural logarithms
r = rate of growth
t = time

The rate of growth (r) is not always 2.0 as it is in bacteria; for example, some fungi can produce more than two daughter cells from each parent cell, but the 'exponential growth law' is equally applicable. It successfully describes phenomena as diverse as the growth of microbial cultures in the industrial fermenters which give us products ranging from antibiotics to organic acids, and the early phases of epidemics of coffee rust or of human influenza. Its importance is that, given measurable, numerical inputs, it will give precise outputs. Thus it provides numerical predictions upon which actions can be planned.

Publication of the four reports did not proceed without some irritation and embarrassment for Harry. As Dyer put it, "... reports to the Royal Society on the Thames Water business were voluminous, and in the aggregate make a stout volume. But the detailed work which he also sent in for publication formed a stack of drawings about a yard high." After some difficult negotiations, Harry agreed to reduce the number of figures, but not a lot: they still occupied 15 quarto plates and would have cost the enormous sum of £400 to publish. To Harry's chagrin, the proposition to publish all of these was lost when it was finally put to the members of the Water Research Committee. Some, such as Edward Frankland, were supportive, several, like Sydney Vines, were non-committal, but others, such as E Ray Lankester, were firmly against (his comments were scathing, describing the figures as "lab. notes, useful only to Ward himself").[20]

The immediate practical benefits deriving from the reports were minimal, but, together with the report of the Royal Commission on Metropolitan Water Supply, 1892–93, they did serve to make the water companies more aware of the need for efficient, regularly monitored filtration. It also gave them the confidence to store water in large reservoirs where sunlight could have its sterilising effect. The incidence of water-borne diseases diminished, though it was not until chlorination of water supplies was introduced early in the twentieth century that water in the cities of Western Europe and North America could really be regarded as 'clean'. Sadly, the situation in other continents, and in parts of eastern Europe, has not yet improved: the United Nations Environment Programme conservatively estimates that 20% of the world's human population still lacks access to safe drinking water, with waterborne diseases killing 15 million children each year.

It is appropriate at this point to consider what Harry's services to the Royal Society or, more particularly, his struggles with the Ginger Beer plant did either to advance his career or to enhance his reputation for posterity. It may be questioned whether, at a critical stage in his career, he lost his natural focus for he took up the Ginger Beer Plant just as he was completing his seminal work on the enzymology of the infection process. He became involved, or rather entangled, in two major projects which had little or no direct connection to plant pathology.

Scientists usually reach their peak between the ages of 35 and 45. Thanks to the happy coincidence of experience and energy, the originality of their ideas and their productivity are then at their highest levels. Judged in terms of modern career development, Harry's strategy, or rather lack of strategy, in determining what work to undertake was risky. He lost focus at a critical stage in his career. He agreed, too readily, to take up requests for his help or to follow Dyer's bidding. Given his propensity for exhausting every detail of a subject, this inevitably led him to commit huge amounts of time to projects that were both diverse in their nature and, it has to be admitted, not always of the greatest importance at the time. In his defence, he applied his intelligence and skills where he thought they would be most useful. In selecting topics to research, he was *applying* his academic knowledge to practical problems and was being true to his own beliefs.

In one sense the strategy worked for he was justifiably well respected by the most eminent botanists of his day, because he always did excellent work, so he gained significant advancement and honors during his lifetime. In another sense it did not. As we will see when his scientific legacy is assessed (chapter 8), the frequent jumps Harry made between disparate subjects – resulting in a loss of focus at the most productive time in his life – cost him the chance of fame much beyond his lifetime. He never made the series of signal advances *in one subject* that would have placed him among the immortals.

Chapter seven: Cambridge – the fatal challenge

"I fully intend to do all that is in me to justify the selection and your opinions, and do not anticipate injury from over work: at least I am not going to be afraid of it, especially as it will be varied, and I realise the value of change of occupation. I believe actual hours of work mean very little so long as life is regular and simple: it is more a matter of generally being at something worth trying. I do not suppose habits will change much, but I do look forward to the new environment, without being altogether dazed by the glamour of it in prospect."

Harry Marshall Ward writing on 10 November 1895, shortly after being appointed to the Chair of Botany in Cambridge University.[01]

On the 21st of March 1895 Harry Marshall Ward reached his 41st birthday. He had by any objective measure made a success of his life. He had the solid home and stable family expected of, and so cherished by, late-Victorian middle-class society. He had climbed the social ladder too, as he had risen in his chosen profession. He had been Professor of Botany at the Royal Indian Engineering College, Cooper's Hill for ten years, during which time he had published some of his finest research. He had steadily gained from his peers recognition which found its expression in his election to Fellowships of the Linnean Society in 1886 – where his supporters included three senior figures at Kew, Thiselton-Dyer, Sir Joseph Hooker and Daniel Oliver (the last was also Professor of Botany at University College, London) – and of the Royal Society of London in 1888. Soon, he was invited to join the Councils of those Societies, their most senior bodies, and in 1893 was awarded the latter's Royal Medal for his work on embryology described in his very first publications. It seems however that Harry was not content and, again, it is Dyer who provides an insight into Harry's state of mind.

A significant mark of the esteem in which Harry was held by his peers had come in 1890 when he was invited to deliver the Royal Society's prestigious Croonian Lecture. He chose as his subject 'The Relations Between Host and Parasite in Certain Epidemic Diseases of Plants'. In the published version,[02] which is extremely short by Harry's standards, we see him defining for virtually the first time a new discipline, physiological plant pathology. He points not just to the intimacy of physiology and pathology but, breaking new ground, he points to the *dynamic* interaction between active host defence mechanisms and products of the potential pathogen. With equally exceptional foresight, he points out how the environment may modify the dynamics of the host-pathogen interaction, for example how the effectiveness of defence might be lowered under sub-optimal growth conditions, as in plants growing in shade. Writing about the Croonian Lecture, Thiselton-Dyer, who predictably saw himself as the person who fixed it for Harry, said it 'put £50 in his pocket'. The latter comment may have been slightly flippant but nevertheless it was made in the context of Dyer's awareness that the smallness of the salary Harry received at Cooper's Hill was becoming an increasing source of grievance. A discontented person often looks for other grievances and, maybe, it was the lack of money which led Harry finally to the opinion, reported by Dyer,[03] that Cooper's Hill was "utilitarian and uncongenial". Undoubtedly restlessness was surfacing in a man who had long nurtured an ambition to be part of a better known, more highly respected institution.

In his Manchester days, Harry had briefly contemplated applying for the vacant Chair at Oxford but was persuaded against doing so because his chances of success were so small. Unsettled at Cooper's Hill, he must have increasingly looked to Cambridge where the death of the incumbent Professor of Botany, Charles Cardale Babington, had long been expected. Quite naturally Cambridge held for Harry a glamor even greater than that of Oxford, since Cambridge was his *alma mater*. Thus, when in the summer of 1895 Babington finally passed away, Harry was mentally prepared and ready to apply for the post. He must have anticipated a stiff fight for what he saw as the greatest prize. To his relief, however, there was little opposition. His most formidable rivals were either already well settled – Vines in Oxford, Bower in Glasgow – or, like Francis Darwin, were not interested. Or, perhaps, Darwin and others like him were just not prepared to undertake the gargantuan task facing the new Professor if Cambridge botany was to be resuscitated. The Chair was offered to Harry who, bravely or blindly or with a mixture of the two approaches, accepted the responsibility. Within eleven years he would be dead.

The problem with Cambridge botany was that it was moribund. Whereas John Stevens Henslow, Chair from 1825 to 1861, had begun to rejuvenate the teaching of botany he soon found other more compelling interests, while CC Babington, Chair from 1861 to 1895, had little interest in teaching. When Francis Darwin became an undergraduate in 1866 he found the teaching of biology (zoology was almost as bad as botany) "in a somewhat dead condition. Indeed, … it had [hardly] advanced much from the state of things which existed in 1828, when my father entered Christ's College".[04] Botany's buildings and facilities were woefully inadequate for either teaching or research. The size of the task facing Harry is best appreciated by examining briefly the personalities and influence of his predecessors.

Henslow's family had suffered financial insolvency,[05] although not as recently as the Marshall Wards. The Henslows were, however, longer established in middle class society and better connected than the Marshall Wards. They were better able to overcome a setback. Young John Stevens was handsome, brilliant, and his rise within the University of Cambridge was rapid. A mathematics graduate and Fellow of St John's College, he was ordained into the Church of England – as was required of all fellows. By the age of only 26 he was appointed to an endowed Chair of Mineralogy and, only three years later, in 1825, he exchanged Mineralogy for the newly vacant Chair of Botany (he actually held both Chairs for two years). Henslow was not a botanist by training but he was linked to the botanical establishment of the day; his wife was connected to the family of Daubeny, the Professor of Botany in Oxford, and Henslow's first daughter, Frances, married William Jackson Hooker, Director of Kew from 1841 to 1865. A position in society and in academia came relatively easily to him. In his early years as Professor of Botany he was an enthusiastic and successful reformer but his university stipend was poor and he was increasingly drawn to the church, in particular to his parish of Hitcham which provided him with a relatively rich living. As described in chapter two, the village school took more and more of his time and in 1839 he finally gave up his house in Cambridge in order to save money. Thereafter, the fate of botany in Cambridge came fairly low in his list of priorities.

CC Babington came from a land-rich family. His education, just like that of his father and uncles, was quite naturally completed at St John's College, Cambridge, where he was a keen and particularly enthusiastic student of Henslow. Indeed, he became Henslow's assistant when, after taking his MA degree in 1833, he became a tutor and later Fellow of his College. After 1839, when Henslow moved to Hitcham, it fell to Babington to keep alive the teaching of botany. It was Babington who led the excursions and taught the 'field botany' once so enthusiastically espoused by Henslow. The cost of Henslow's increasingly prolonged absences soon became apparent. Whereas he had had classes of between 70 and 80 students (including Charles Darwin) in the first five years of his professorship, numbers had fallen to 10–20 in the 1840s, when Cambridge botany reached perhaps its lowest ebb. Numbers only revived after changes in the university regulations opened the way for the new Natural Sciences Tripos of 1851; this immediately made botany a more popular subject in its own right since, for the first time, it allowed students who had a strong interest in the subject to take a Pass degree and, from 1860, an Honours degree in *botany*.

The 'New Botany' was a late arrival in Cambridge. Although as early as 1835 Henslow had published a book entitled *Principles of Descriptive and Physiological Botany*, its physiological content was neither original nor up to date for it relied heavily on the largely outdated writings of the French botanist, de Candolle. After the Irish Potato famine of the 1840s, Henslow developed an interest in plant pathology, but neither he, nor Babington, was in any way an experimentalist. As Max Walters has pointed out in his *The Shaping of Cambridge Botany*, "neither Henslow nor Babington ventured outside the British Isles for more than brief periods, and neither, apparently, could converse in any modern foreign language except a little French".[06] The price of such a deficiency became costly after Babington finally assumed the chair of botany on Henslow's death in 1861 for he was effectively isolated from the great explosion of botanical knowledge coming out of Germany (chapter 2). The most positive legacy of the Henslow/Babington era was the New Botanic Garden, which was opened in 1846 at an extensive site on the south-east borders of the city. It was Henslow's vision that plants from Europe and around the world would be collected and grown in such a garden where they would provide living materials for the attention of taxonomic researchers and students alike. He managed to persuade the university to support the Garden although he made little use of it himself. For his part, Babington was always more comfortable in the Herbarium and in his later years tended to regard the Garden as nothing but a nuisance.

Babington was already 53 years old when he became Professor but he held the chair for another 34 years, in an increasingly feeble state of health. For the last four years he was in official semi-retirement. His inactivity and withdrawal from his department made life increasingly difficult for his staff. None suffered more than Sydney Vines and Francis Darwin.

We have seen in chapter 2 how Vines' early attempts to teach his students a little of the physiology emerging from the German laboratories succeeded only because of the generous provision of teaching space by Michael Foster, Professor of Physiology. It was not until 1881 that Babington allocated space to Vines, in a small annexe to the Herbarium, where he could accommodate about a dozen workers with their microscopes. His teaching load was heavy, and made heavier in 1882 with the transfer of The Elementary Course to him when Hillhouse, previously assistant Curator in the Herbarium, left to take the Chair at Mason's College, Birmingham. Vines was to all intents and purposes the head of botany but Babington, who had almost totally ceased to give lectures, lived on. So, in 1883, the university created for Vines a new post, that of Reader. Vines' reputation was attracting increasing numbers of students and it became necessary for the university to add a new building alongside the Herbarium to accommodate up to 100 students. With an increased workload Vines became a victim of his own success, and it seems there was some internal friction within the university.[07] Not surprisingly, when the Sherardian Chair of Botany at Oxford became vacant in 1888, and with Babington clinging on, Vines judged that enough was enough and left for Oxford. It must have been a very difficult decision for him to leave his own university for 'the other place'. Sadly, he was a spent force. His best years were past, and he soon became something of a recluse in his home on Headington Hill, on the edge of Oxford, much to the dismay of his new colleagues in the university.

Francis Darwin (Fig. 7:1) was appointed University Lecturer in 1884. A physiologist of considerable distinction, whose first publications on the tropic movements of plants were made jointly with his father, Charles, Francis helped share Vines' burden of teaching and administration. He was made Reader on Vines' departure and in 1890 he astutely added at modest expense a 'house-property' to the department's facilities. It was not in a good state of repair and its internal design was far from ideal, but it was conveniently located almost next to the old buildings. In 1891 it was finally recognised that Babington's decline was irreversible and Darwin was appointed Deputy

Fig. 7:1. Francis Darwin FRS: a self-effacing man who was a tireless biographer of his father. In spite of his reputation for preferring dogs to men, he was generous in thought and deed. (By permission of the Royal Society.)

Professor. Adding to Darwin's problems was Walter Gardiner, another member of staff in decline, at least academically. Much had been expected of Gardiner (a fellow student of Harry's at Cambridge), who was the first person to demonstrate the cytoplasmic strands, called plasmodesmata, which connect the cytoplasm of many types of neighboring plant cells. Gardiner's interest in the department, except for its Museum, had sadly fallen away. He became more and more immersed in Clare College, where he was Bursar, and less and less of help to Darwin. Dyer wrote that when Babington's Chair finally became vacant, Darwin 'generously waived aside his claims' in favor of Marshall Ward.[08] In truth, Darwin was finding the demands of work were conflicting with his growing interests in music and literature. Also, unlike Harry, he did not need more money for he had a private income on which he could live comfortably (Darwin's obituarist noted that his hours of leisure were not curtailed through the necessity of adding to income by any form of drudgery). Above all, however, he did not want the Chair because he was probably all too aware of the scale of the problems awaiting the incoming Professor. Darwin proved a critically important supporter of his old friend Harry Marshall Ward's application for the position. Darwin knew well the energy, abilities and raw ambition of the younger man, what he could not have known of was Harry's fragile health.

Apart from Darwin, who was to be further distracted by the death of his second wife in 1903, and Gardiner, who was embittered because he had been ruled out from applying for the Chair on grounds of ill health,[09] Harry inherited a staff comprised of AC Seward, a palaeobotanist, and the young FF Blackman, a recently appointed physiologist. He inherited buildings that were cramped and ill suited to either teaching or research. As the quotation[01] at the start of this chapter intimates, even as he took up the post, the glamor of the ancient university did not blind him to the reality of the difficulties facing him as he attempted to modernise its teaching and to establish a professional research culture. For this is what Harry was – a professional. For him, botany could no longer be a gentleman's hobby. It would have an important contribution to make to the wealth of the nation and to the health of its people in the new century, which was only five years away, but to do that it needed a supply of rigorously trained professionals. Everything Harry achieved at Cambridge stemmed from his early recognition that having the best modern laboratories was the key to delivering the best modern training; just as they were to providing an environment in which research could flourish. He dedicated himself to a campaign to get those facilities.

The struggle was long and no doubt hard but by 1st March 1904 there stood on the Downing Street site a set of new science buildings ready to be opened by King Edward VII (Fig. 7:2). Among those buildings was a new Botanical Laboratory (subsequently known as the Botany School) designed by the architects Marshall and Vickers but very much according to Harry's instructions (Fig. 7:3). It was widely recognised that the presence of a building for botany owed everything to the extraordinary persuasive and administrative skills of the Professor of Botany.[07] The ground floor of the simple rectangular building was unexciting, incorporating a large lecture room for 200 students, herbarium and museum, but the first and second floors broke with tradition. On the first, the principle was established that each member of staff should have his own research laboratory, in addition to which there were laboratories for the study of morphology and chemistry, and a library. On the second, the principle was continued, in addition to which there was a laboratory for physiology and a glasshouse. The crowning glory, however, was a large Elementary Laboratory for the teaching of up to 150 undergraduates (Fig. 7:4). Each room was well equipped and there was space for everyone, including two juniors, University Demonstrators, who had joined the staff by 1904. Even the flat roof was used as the base for further glasshouses. The professor's own suite was modest by the standards of the day but, typically, was exceptionally well thought out. In addition to his office, laboratory and own small glasshouse, he had rooms for the incubators and

Fig. 7:2. Several local newspapers produced commemorative issues to mark the historic Royal Visit. (By permission, Cambridgeshire Collection, Cambridge Central Library.)

sterilisers which by then had become an essential part of plant pathological research.

In these buildings Harry could, at last, teach what he thought was important, using the methods that had first impressed him so much in Germany some 20 years earlier. He led by example, taking on a major part of the elementary teaching. 'The Professor's General Course' consisted in the first term of the morphology and anatomy, plus some physiology, of flowering plants, followed in the second by a course on fungi, algae, bryophytes, vascular cryptogams and conifers taught from an evolutionary standpoint. No doubt the Professor drilled into his students again and again the importance of relying upon their own eyes, and not on textbook diagrams, and the principle that drawing and labelling is the best way of committing observations to

Fig. 7:3. The Botany School, designed by the London Architects Marshall and Vickers to Harry Marshall Ward's detailed instructions. The architect WC Marshall was not related to Marshall Ward but was a friend of Francis Darwin. (By permission, Cambridgeshire Collection, Cambridge Central Library).

Fig. 7:4. A laboratory in the new Botany School. (By permission, Cambridgeshire Collection, Cambridge Central Library.)

memory. In the third term the emphasis was on the systematics of flowering plants, with visits being made to the Botanic Garden. In the long summer vacation there were extended field trips and excursions.

Not surprisingly, it would be to the Elementary Laboratory that Harry led the King and Queen when they came to open the new building. That day was probably the proudest of his life. It represented the culmination of all his ambitions. Here was recognition. He was a Professor in one of the most respected universities in the world, surrounded by all its pomp and its dignitaries. He was showing to the Royal Family the Botanical Laboratory which he had, almost literally, built himself. Later in the day he would entertain his staff and also guests who included some of the most eminent botanists of the day. His health would soon decline and in two years he would be dead, but for the moment let us, like Harry, savor the day.

A light dusting of snow had fallen overnight but by midday, when the King's train arrived from London, it had been largely thawed by bright sunshine, albeit accompanied by a biting East-Anglian wind.[10] The Queen and Princess Victoria had arrived ten minutes earlier from Sandringham; after brief speeches from Cambridgeshire Councillors and the King, the royal party's carriage took them through crowded and decorated streets to the Senate House (Fig. 7:5). Here, in the presence of the Professors and other senior members of the University, the Vice-chancellor welcomed their majesties and outlined the purpose of the day, to open new buildings; the Squire Law Library, the Medical School, the Botanical School and the Sedgwick Museum (Geology). Winning prolonged cheers from the rowdy undergraduates in the gallery, the King tactfully started his reply, "I am

Fig. 7:5. King Edward VII's carriage arrives at the University Senate House. (By permission, Cambridgeshire Collection, Cambridge Central Library.)

proud to be a member of the University, and I shall always look back with great pleasure upon the time which I spent as an undergraduate at Trinity". (A few years earlier, when he was Prince of Wales, 'Bertie' had earned himself a reputation as a particularly boisterous undergraduate, not to say a hell-raiser.) Continuing, the King told his audience, "the extension and development of all branches of study and research … are essential to the maintenance of the greatness and the welfare of my Empire". Again tact-fully, he pointed to the special responsibilities that fell upon the older universities in this respect.

From the Senate House, the royal party moved on to a seven-course lunch at the Fitzwilliam Museum, at which Harry was among the few members of the university privileged to be present. There was also a short guest list, which included Sir William Thiselton-Dyer and Sir Archibald Geicke, both of whom taught Harry at South Kensington and were possibly on the list at Harry's suggestion. After lunch came the most fulfilling but nerve-racking part of the day, the guided tours. Led into the Botanical School by Harry and Francis Darwin, the royal party inspected the Museum and then, as the *Cambridge Review* reported, "courageously climbed a long staircase to the Students' Laboratory where, by happy thought, a number of men were at work". The King took great interest in the operations of the students, to some of whom he spoke.

The visit was not without its embarrassments. First, without forewarning and for reasons unknown, the King expressed a wish to be introduced to Walter Gardiner. After a frantic search, "Mr Gardiner was found not to be in the building". He was, in fact, walking on the Gogs, nearby hills.[11] His absence was all the more remarkable because he was devoted to Queen Alexandra and each year on her birthday made the journey to London to present her with a bouquet of flowers from the University Garden. His behaviour could be explained by his increasing eccentricity but, perhaps, he just could not bear to share his Queen with Professor Marshall Ward, the man who occupied the Chair he had coveted so much. The second embar-rassment was that, in the heat of the moment, Harry failed to get all members of the royal party to sign the Visitors' Book, which was to be the permanent record of the great day. All too soon, but perhaps not soon enough in view of Harry's nerves, and to the accompaniment of cheering, the party moved on to the Adam Sedgwick Museum and the unveiling of a statue to the late Professor.

And then the less formal celebrations could begin. Harry and Selina took the most unusual step of entertaining staff and guests. Although called an 'At Home', the event took place in the Botanical Laboratory where, without the restrictions of modern Health and Safety Regulations, both food and drink could be enjoyed (Fig. 7:6). The function represented a way in which Harry could thank his staff for their support during the trying years through which they had all passed. His mind had often been pre-occupied by the problems of the building and, as Dyer wrote,[12] "In later years, in conversation especially, thought seemed to come quicker than words. … In such a predicament he

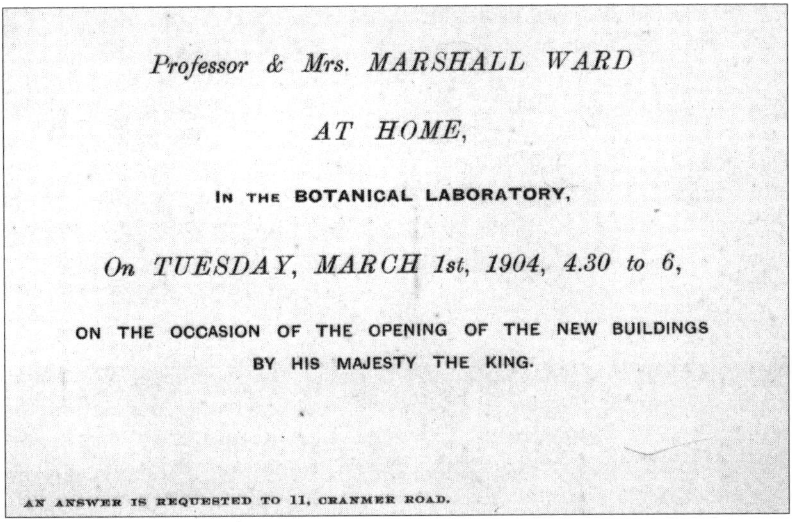

Professor & Mrs. MARSHALL WARD

AT HOME,

IN THE BOTANICAL LABORATORY,

On TUESDAY, MARCH 1st, 1904, 4.30 to 6,

ON THE OCCASION OF THE OPENING OF THE NEW BUILDINGS
BY HIS MAJESTY THE KING.

AN ANSWER IS REQUESTED TO 11, CRANMER ROAD.

Fig. 7:6. A card offers an invitation to the final event of a memorable day. (By permission, the Royal Botanic Gardens, Kew.)

would simply remain silent, and slowly move his head. This habit, I think, explains the reputation of being 'mysterious' which he seems to have acquired latterly in Cambridge."

It seems that on that bitterly cold spring evening in 1904 he was trying as best he could to acknowledge that he had been a difficult man to work with and to make amends. For this brief moment, the immediate worries of the past could be put behind him and a brighter future anticipated. He could relax in the warm glow of achievement. It is reported that he was, quite properly, the recipient of "many hearty congratulations"[13] and, no doubt, an equal number of toasts.

The 1st March 1904 cannot be left without some mention of the events that occurred on the same evening, since the extensive newspapers reports of the events remind us how similar are Harry's and our own age. They show how, in any age, young men in a gang or a mob, and fuelled by alcohol, can go wild, casting aside their normal respect for property and the law. There is a familiar ring to *The Cambridge Express*' account[10] of the way events unfolded. Ill-feeling had grown all day between the police and the huge numbers of people who had crowded the centre of Cambridge trying to catch a glimpse of the King and his retinue. The local police force was small and, to help them cope with the crowds, a number of Metropolitan policemen, some on horseback, had been brought in from London. There had been much pushing and jostling, and as tempers rose "the Metropolitan policemen … certainly did not err on the side of handling the crowds too gently". Later in the day, firecrackers were thrown at the police horses, and the helmets of several policemen were either forced down over their ears or knocked off and used as footballs. Towards nightfall, the crowds were swelled by undergraduates, fresh from their evening meals and drinks, and soon the young gentlemen of the university engaged the equally young belligerents of the town in their age-old 'gown versus town' battles. The mob tried to light a fire in the centre of the city but was finally driven away by the police and onto a large green space known as Parker's Piece. There, bonfires were built with palings torn from fences, shutters from windows, notice boards – a dentist and an auctioneer lost their professional signs – and a watchman's hut. It was late into the night before order was restored. Local courts were busy for several days dealing with the long queue of law breakers that the police had managed to get their hands on.

Harry Marshall Ward was never able to enjoy the relatively relaxed lifestyle of Henslow or Babington. Owing both to his personal inclination to do his duty and to be involved, and to the fact the university was growing and becoming more bureaucratic, Harry's time was from his arrival in Cambridge heavily committed to meetings and committee work. As if this and all the work entailed in winning approval and finance for the new Botanical Laboratory, and then in its design and fitting, were not enough, the British Association (BA) was making increasing demands upon his time. A long-time member and staunch supporter of the BA he had, ever since he gave his first paper at the Southport meeting in 1883 (on fatty bodies in the cells of coffee), found its annual meetings the perfect platform for publicising his research. The meetings were also, as we have seen in chapter four, an ideal place to meet old and new friends. Through both the BA and the Royal Society, he became known not just to botanists but to the leading zoologists, chemists and physicists of the day. To his satisfaction he was increasingly frequently asked to join committees of the BA and to play a leading part in its administration. It helped him to cement his place among that small group of his peers who formed the scientific establishment of the day.

Meetings of the BA covered the whole spectrum of the sciences but quite naturally, given the diverse topics that its meetings included, the programmes were divided into sections covering different subjects, such as Physics, and Geology. For its part, Biology had always been treated in one section, so it was a mark of the coming-of-age of botany, of how far Harry and the men of his generation had taken the subject, that in 1895 the BA created a new section, K, specifically for botany. The change was timely for Harry since when the BA decided on the adventurous step of taking the 1897 meeting across the Atlantic to Toronto, Harry was invited to be President of the newly-fledged Section K.

The invitation was both an honour and an opportunity to travel to a continent of which he had only read. At first he had hoped Selina would accompany him on what would be a great adventure but, as he wrote to his friend Will Farlow,[14] "Mrs Ward cannot leave our two children, and is moreover too fearful of the sea voyage so that I am sorry to say she cannot come with me". Sailing alone on the SS 'State of California', Harry arrived in Quebec on 30 July, having allowed himself almost three weeks for travelling before the conference began. He had hoped to visit Farlow in Harvard and, possibly, take in a meeting of the American Association, but whether these plans materialised is not known. Certainly, Farlow attended the BA meeting in Toronto, staying at the same Queen's Hotel as Harry, but the records suggest he did not register until 21 August. If that were the case he might have missed Harry's paper on *Stereum hirsutum*, a wood-destroying fungus, given on 19 August and his Presidential Address to Section K given on 20 August. Perhaps he knew what the latter was to entail!

Harry loved the history of his subject and had an encyclopaedic knowledge of its major players and their roles, but brevity was never his strength. Sydney Vines was to reflect after Harry's death, "… he generally attempted to cover a great deal more ground, and to convey a great deal more information in his lectures, than was possible either physically or mentally".[15] Never was this more true than in Toronto; the written version of his Address on the history of mycology and plant pathology runs to 28 pages and is almost three times longer than that of any other sectional President's address. If the spoken version matched the written, it would have filled most of the time between its start at 11.30 a.m. and the next timetabled talk at 2.00 p.m. It might have won him reluctant admirers but it would not have won him many new friends.

The organisers of the meeting arranged a two-day excursion to Niagara to see the Falls and the remarkable new single-arch steel bridge over the Niagara River which was to be opened the next month (Fig. 7:7). Harry took this trip into the mists and rainbows and on 27 August, when the conference was ended, he joined a number of 'B asses' (as Raphael Meldola the pioneer ecologist called them in his memoirs[16]) on a train that crossed the continent. After crossing Lakes Superior and Huron by boat and stopping in the prairies and Rockies to make various excursions to see the local flora and fauna, not to mention the 'red indians', the party finally reached Vancouver Island. On the return journey, Harry and a small group of his friends were joined by Professor Fernow of the US Department of Forests. Again, travel was punctuated by stops. Harry took the opportunity to collect seeds, while his old friend Edward Poulton collected insects. It seems that their local guide, Fernow, suffered quite a serious fall while the group was exploring in the Cascade mountains but, thankfully, he and Poulton, who was slightly injured in the same incident, recovered quickly. In some perfectly dreadful verses sent to the SS 'Parisian', as the Englishmen sailed home, Fernow wrote[17]:

> "Don't slide down gangways, for it has been proved
> – The painful proof is mine –
> There are degrees (of Downward grade)
> Professors even should decline …"

Whereas the Toronto meeting and the whole North American expedition of 1897 were great stimuli, leading Harry to take a rare and extended holiday in

Fig. 7:7. During and after the 1904 British Association meeting in Toronto, Canada, a number of excursions were organised for delegates; one was to the nearby Niagara Falls. (MS. Dep. B.A.A.S. 186, by permission The Bodleian Library, University of Oxford.)

which he could once again indulge his almost forgotten interest in plants, far away from the cares of Cambridge, the 1904 meeting of the BA was the exact opposite. Clearly, Harry and the other science professors wanted to show to the rest of the world their wonderful new lecture rooms and laboratories so recently opened by the King. But for Harry in particular the extra workload generated by the meeting, coming on top of that associated with the opening of the buildings, seriously eroded his health. He was not President of Section K for the Cambridge meeting – Francis Darwin fulfilled that role – but he had the responsibilities of a Vice-President. In addition, he was involved in at least three committees whose work had to be completed in time for the meeting. He chaired a committee on 'Experimental Studies in the Physiology of Heredity', the work of which was to report on ground-breaking studies on the nature of genetic variation in *Primula* being made by William Bateson. He chaired another committee on 'The Respiration of Plants', which reported on the use of small grants made by the BA to FF Blackman to enable him to purchase apparatus needed for his investigations of the effects of temperature and light on photosynthesis and respiration in *Helianthus* (see chapter 8). Harry was also a member of a large committee, whose other members included Sir Michael Foster, Sir Oliver Lodge and the Bishop of Hereford, looking at 'The Influence of Examinations'. This last committee looked at the ways in which schools in England and Wales prepared candidates for the entrance examinations for universities. This huge topic, not surprisingly, required the collection of an enormous amount of data, some of which was for the sake of comparison drawn from Scotland and the USA. It generated a substantial final written report whose preparation, no doubt, cost the committee members much time and effort. To cap his other commitments to the Cambridge meeting, Harry decided to use the 45 minutes allocated to him in the scientific programme of Section K to deliver a rare attack on a fellow plant pathologist. An extended form of the paper was published in 1905 as a major review,[18] 'Recent Researches on the Parasitism of Fungi'. It was his last publication.

When his time in Cambridge had permitted Harry to continue with personal research, he had again changed focus but, this time, he had gone back to a subject where he had already established a reputation – to the life histories of rust fungi. He turned not to the rust that attacks coffee bushes but to those that attack grasses, including cereals. The most recent authority on the subject was the Swede, Jakob Eriksson. Harry soon found himself disputing some of Eriksson's basic ideas; his opposition was both forthright and, ultimately, very public.

Eriksson was puzzled by infections which arose where spores showed no visible connection to fungal mycelium inside a leaf and, more particularly, by the those sudden outbursts of rust infection that occurred in spring in the apparent absence of an obvious source of either winter spores from infected cereal tissues (living or dead), or of other rust spore types deriving from a secondary host plant species, for example, barberry in the case of wheat stem rust (Fig. 7:8). He accounted for these phenomena with his celebrated theory of the Mycoplasm launched in 1897. The gist of this implausible theory was that the naked protoplasm of the pathogen lives for a time mingled in indistinguishable combination with that of the host. This mixed plasm was the Mycoplasm. At certain stages in host development, the pathogen protoplasm separated itself and migrated across the cell wall outwards into the

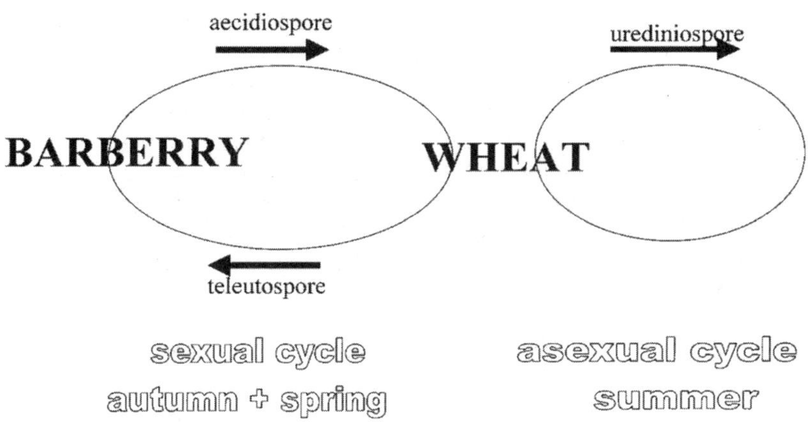

Fig. 7:8. Life cycle of wheat stem rust, *Puccinia graminis* f.sp. *tritici*, showing three of the five spore types. In summer the fungus spreads rapidly from wheat plant to wheat plant producing asexually several generations of urediniospores. In mild climates this cycle may continue over winter. In harsher climates, teleutospores are produced in autumn. These tough spores overwinter and in spring give rise (indirectly via two further spore types) to infection of the secondary host, barberry. After sexual reproduction in barberry, aecidiospores carry the infection back to wheat.

intercellular space. It then became thread-like, secreted a cell wall and sent absorptive organs, called haustoria, into adjacent cells. Mycelia then developed and the fungus emerged from the host to sporulate. Stretching credibility even further, it was assumed the pathogen then managed to infect seeds, reverting to the mycoplasmic condition and ensuring its own transmission to the next generation of the host.

Harry was almost certainly sympathetic to the difficulties of tracing fungal hyphae in infected plant tissues. He might have recalled the letters he wrote from Ceylon to Thiselton Dyer in which he explained the difficulty of finding the physical connection between the spore on the surface of the leaf and the mycelium inside it. Thus, when first dealing with the Mycoplasm theory, at the time of his address to the British Association meeting in Toronto (1897), he made an almost even-handed presentation. "This is a startling hypothesis, and brings us to the most advanced point along this line of biological speculation. We must distinguish sharply and clearly between such a view, which is by no means inconsistent with all we know of parasites, so far as dormant mycelium goes, and all the hazy, mystical suggestions as to 'infective substance' and so forth, which were so freely flung about at the beginning of this epoch, and which De Bary's strictly scientific methods put down so firmly."

Once he began his own meticulous microscopical studies, the time for even-handedness was over. Studying rust on *Bromus* spp., just as Eriksson had done, he soon saw that spores on the leaf surface made connections with interior of the leaf via exceptionally fine hyphae. These were easily missed unless they were observed in sections cut vertically through the leaf so that the long axis of each thread was exposed. As each spore germinated on the leaf surface, the germination tube formed a swelling over a stomatal pore on the leaf and from this developed an infection thread. Once within the sub-stomatal chamber, the thread developed into a swelling that, in turn, led to infection of the leaf's cells. Inside those mesophyll cells were not the mysterious 'special corpuscules' described by Eriksson but the absorptive organs, called haustoria, he had previously seen in infected coffee leaves. "Eriksson has entirely reversed the true order of events" stated Harry in a paper read to the Royal Society in 1903, and published in its Philosophical Transactions[19]. EM Freeman, a young American visiting Harry's laboratory for a year, and studying rust on *Lolium temulentum*, came to the same conclusion. In addition, Freeman found that rust mycelium could survive over winter by infecting the seed of *Lolium* during its development in autumn. Seed transmission is unusual in rust fungi (though normal in smut fungi) but instances of its occurrence may have misled Eriksson towards proposing his Mycoplasm Theory.

And so to the Cambridge meeting of the BA, mentioned already, where Ward proceeded to destroy the Mycoplam Theory using, in large part, ammunition assembled by Freeman. As Vines remarked[15] diplomatically after Harry's death, "One of the most interesting discussions ... was that in which the *pros* and *cons* of this [Mycoplasm] theory were urged by the two protagonists". Balfour, who had also witnessed the event, remarked[20] more bluntly that the outcome "was not to the advantage of Eriksson". By the time reports had filtered across the Atlantic to HH Whetzel, a respected chronicler of early plant pathology, the clear impression was left that Harry's attack had been withering.[21] It is to Harry's credit that, as an organiser of the Cambridge meeting and a man of influence with the Editors of the *Annals of Botany*, he was prepared to give his opponent an opportunity to present his own case. Eriksson was given the next slot after Harry's on the BA programme (though of 20 minutes duration rather than Harry's 45), and their written papers were subsequently published side by side in the *Annals* (1902).[18]

In this period of research in the early 1890s, several fundamental ideas of lasting value emerged from Harry's studies of rusts. Studying *Uredo dispersa* (now *Puccinia recondita* f.sp. *brominia*) on brome grasses, he concluded that predisposition to, or immunity from, infection of the host was not determined, as many thought, by anatomical differences such as size and number of stomata, nor by hairs on the leaf surface. Nor was predisposition affected when he altered the mineral nutrition of the plant, starving it of phosphorus, nitrogen, calcium or other mineral elements. These conclusions we still hold to be fundamentally true today although we would use the terms susceptibility and resistance in place of, respectively, predisposition and immunity. (It is believed today that anatomical or environmental features can have direct 'predisposing' effects on the plant. However their effects are minor, quantitative ones affecting the speed and/ or extent of infection rather than qualitative ones affecting susceptibility, that is whether or not the pathogen is able to complete its life cycle on the host.) And so the anatomical hypothesis had to be abandoned. In the view of Whetzel, Ward "laid the foundation for all later investigations into the nature of susceptibility and immunity in plants".[21]

There was, however, one area of Harry's studies of the ways in which rust fungi interact with plants where his conclusions were completely wrong. It is ironic that in this instance Harry was supporting Eriksson and his 'Bridging Theory'. In 1894 Eriksson had proposed, correctly, that there could exist within one species of a pathogenic fungus a number of morphologically similar, or indistinguishable forms, each with different host species. He also proposed, incorrectly, that these forms were not absolutely fixed. Thus, for example, the form of stem rust that normally attacked wheat might adapt to attack barley and then, in turn, rye. Barley would thus be a 'bridging species'. Eriksson was supported in his opinion not only by Harry but by a number of other authorities of the day.

It was not for almost another 20 years that the theory was finally, and comprehensively, destroyed by a young American pathologist, EC Stakman.[22] Not only did he point out its lack of logic – how could there exist several biologic forms distinct from one another if bridging occurred? – but he carried out an exhaustive series of experiments which produced not a shred of evidence in support of bridging. Beginning with his PhD thesis, *A Study in Cereal Rusts – Physiologic Races* (1913), Stakman clearly demonstrated

		PATHOGEN						
		Physiologic races						
		1	2	3	4	5	6	7
	Group a	+	+	+	–	+	–	–
HOST	Group b	+	+	–	+	–	+	–
CULTIVARS	Group c	+	–	+	+	–	–	+

- + and – show, respectively, susceptible and resistant host reactions
- within a group there may be any number of cultivars; in practice, each group is represented by a 'differential' cultivar, a type-specimen
- the number of races that can be identified is equal to 2^{n-1}, where n is the number of cultivar groups: in this case n = 3, so 7 races can be identified
- when a fungus is isolated from the field (each culture is normally grown from a single spore) each isolate will be identified as belonging to one of the races according to the pattern of reactions it evokes in the 'differential' cultivars.

Fig.7:9. A schema for the identification of physiologic races within a species, or *forma*, of a plant pathogenic fungus. Each new isolate may be assigned to a race but not all of the theoretical races may yet have evolved in practice.

that the rust species *Puccinia graminis* can have several special forms (formae speciales = f.sp.) each attacking, for example, only wheat, or oats, or rye (called, respectively, *Puccinia graminis* f.sp. *tritici*, or *P.graminis* f.sp. *avenae* or *P.gramins* f.sp. *secalis*) but not the other two cereals. He went on to show that within each forma there is the potential for a huge number of 'physiologic races' to occur, each extant race capable of being distinguished on the basis of its ability to attack some cultivars, or varieties, of the host but not others (Fig.7:9). The cultivars used to distinguish the races were called 'differentials'. Subsequently, for over half a century, race analysis in this style proved a major tool in combating stem rust of wheat (*P. graminis tritici*), and countless other major crop pathogens which have a comparable structure of *formae* and races, because for each crop it allowed breeding to be targeted at those fungus races most prevalent at a particular time or in a particular location.

Where had Harry, like so many others, gone wrong? With hindsight we can see that he probably obtained misleading results because he had not used single strains of the fungus. As noted earlier, rust fungi can be cultured only on living host plants. So, although we can now store their spores deep-frozen or in a vacuum, the growth of genetically distinct *formae speciales* and physiologic races still requires, just as in Harry's day, living host seedlings which must be kept in physical isolation from each other. This is extremely demanding of both space and time and it is all too easy to mix fungal strains – to cross-contaminate the host plants – which was probably what Harry did. Of course, pointing to such difficulties merely underlines the excellence of Stakman's work.

Despite this isolated failure, Harry's research made considerable progress during his Cambridge years. He achieved this while revitalising the Botany

School, securing for it new buildings and injecting into it a new spirit in teaching and research. But the strain was immense. He became a detached, uncommunicative figure, far removed from the young man who so happily sang for his supper to the Governor and polite society of Ceylon. The lean and austere features of the Cambridge portrait (Fig. 7.10) are those of an overworked man whose thread of life was nearing its end.

Fig. 7:10. Harry Marshall Ward near the end of his life. The portrait hangs in the Department of Plant Sciences, University of Cambridge. (By permission of JC Gray for the University of Cambridge.)

Chapter eight: Legacies

"What we used laughingly to call 'The Cause' – the establishment of a vigorous botanical school in this country, independent of Germany – possessed all his energies. To this must be added a noble and uncomplaining courage which enabled him to face straitened circumstances and declining health with unfailing cheerfulness."

Sir William Thiselton-Dyer (1907) in his obituary of Harry Marshall Ward, written for *New Phytologist*.[01]

Of all Harry Marshall Ward's scientific colleagues and friends, Dyer had known him longest and was best placed to recognise the impending tragedy. At a meeting of the Royal Society in 1905 Dyer saw all too clearly, "The hand of death was obviously upon him: he was wasted to a skeleton, and his eyes shone with an unearthly glitter. I implored him to take a long leave of absence, and a rest in a warm climate. He assured me, however, he would soon be well".[01] By the late summer of 1905 it was clear even to Harry that his health was deteriorating rapidly. Nevertheless, he insisted on carrying out his commitment to represent the Council of the Royal Society at the 2nd International Botanical Congress in Vienna.

Harry had the classic symptoms of what would now be recognised as Type-1 diabetes but we do not know whether an accurate diagnosis was made before his death. In any case, there was at the time no hope for him for, despite the link between diabetes and degeneration of the pancreas being widely accepted by 1906 following the evidence of numerous autopsies, it was not until 1921 that effective treatment with insulin was announced by two young Canadians, Frederick Banting and Charles Best. On his way back from Vienna, Harry visited two famous German clinics. First, he spent three weeks at Carslbad but, receiving no benefit, he went on to Dr von Noorden at Sachsenhausen, near Frankfurt. Nothing could be done for him and he was advised to return to England in easy stages.

He fought on through 1906, even talking to his family about making a return trip to the tropics. In July his son, Frank, happily left home for a holiday in the Tyrol and Selina took Harry to Devon, the county where she had grown up and where she hoped the balmy sea air of Torquay might help him. Tragically, he was past help; he quickly became much weaker, finally falling into a coma and dying on the night of Sunday 26 August. Frank could not be contacted in time to reach his father's bedside but, thankfully,

Zachary, the oldest of Selina's brothers, was able to be present and to support his sister in her grief.

Harry's death certificate states 'Diabetes mellitus 3 years' as the cause of death, the final six months having been marked by 'Pthisis', a general deterioration in the condition of his lungs. Three years might or might not have been an accurate estimate of the onset of the disease but it points to a long struggle for Harry against a progressive, energy-sapping tiredness. It was a cruel end for a man who was normally so energetic and fully committed to life. In addition to a progressive loss of weight, his suffering would probably have been intensified by blurred vision, dry skin, and bouts of intense thirst or hunger.

Harry's body was returned to Cambridge for burial on Monday, 3 September. Frank was still travelling back from the Tyrol, but by Selina's side at the funeral were her daughter Winifred, father-in-law Francis, brothers Zachary and James, and brother-in-law Stephen Glanville. The *Cambridge Independent Press* of 7 September 1906 reported the event but made no mention of Harry's sisters being present. However, the University, City of Cambridge, Linnean Society, Royal Botanic Gardens, Kew, and the Royal Society were all represented. Both Vines and Thiselton-Dyer were present. Among Harry's colleagues, AC Seward and Walter Gardiner were present, as was his research student FT Brooks.

The net value of Harry's estate was £7,239 9s 11d, a not insignificant amount of money for the times. The problem for Selina was that the money was to be the sole support for herself and her two adult children for she had no widow's pension. To make matters worse, her fine seven-bedroomed home at 11 Cranmer Road, a broad leafy avenue on the edge of the city and known locally as 'Don's Alley', was rented from Trinity College. She had no equity to draw upon.

It was immediately clear that 20-year-old Frank could no longer continue the leisurely pace of his studies at Christ's (his father's old college in Cambridge), so he accelerated them, gaining a second class honours degree in two rather than the usual three years. At this point he received help from some of his father's old friends. First, the Professor of Chinese told him about a teaching post available at the Shanghai Public School. Frank was desperate to travel to the East, the Professor pulled some strings, and in January 1907 Frank left Britain to take up his new duties in China. He spent the long school holidays travelling in the Dutch East Indies and briefly joined an American zoological expedition to northern China, but he may well have remained teaching in Shanghai if it had not been for another family friend, Balfour – by now Sir Isaac Bayley-Balfour. From time to time, Sir Isaac had advised Arthur Kilpin Bulley, the wealthy Cheshire cotton merchant who was the founder of 'Bee's Seeds', on what and from where exotic plants might be imported from the tropics. When in 1910 Bulley's favorite plant hunter, George Forrest, was poached by another firm and Balfour was asked to suggest a replacement, he remembered that Frank was a keen and accomplished botanist and was already in the East, able at short notice to mount a new expedition through Burma to Tibet. Frank was appointed and early in 1911 began his legendary exploits. It was on this very first trip that he saw for the first time the plant with which his name has become associated, the blue poppy (*Meconopsis betonicifolia*). It is fitting that the first of his many books, *The Land of the Blue Poppy* (1913), was "affectionately dedicated" to the memory of his father from whom he inherited his love of both plants and the tropics.[02]

While the prospect of Frank completing his studies helped Selina's financial plight, other measures were required to conserve the money Harry had left his family and the most obvious was to give up the large house in Cranmer Road. The exact timing of events is not clear but in October and November of 1906 she was addressing letters from Highgate, North London, where she, and presumably Winifred, stayed with Selina's younger sister, Nannie, and her husband, Stephen Glanville. He was a journalist and, unusually for pre-Great War Britain, she worked – as a district nurse – suggesting the couple were probably exceptionally open-minded and progressive. It may well have been they who helped persuade Winifred of the need to think of a future career for herself. She did indeed succeed in building a professional career, as a speech therapist – one of the first women in this new profession. She never married, remaining devoted to her family and, in particular, to her brave and adventurous brother.

It is not known how long Selina and Winifred stayed in Highgate but *Spalding's Directory of Cambridge for 1907* records Alfred North Whitehead was living at 11 Cranmer Road, when there is no record of Mrs Marshall Ward anywhere in the city. However, by the time the next *Directory* was published in 1910, she was living at 23 Chesterton Road, a small terraced house close to the river Cam.

For a time, Selina managed to keep her mind occupied by tidying up the many unfinished projects left behind by Harry. She had often helped him by proof reading his forthcoming publications and generally liaising with his publishers. Now, in the aftermath of his death, and with the help of Sir Joseph Hooker, she was able to busy herself with his last great project, volume 5, entitled 'Trees', in a massive series, *A Handbook of Forest Botany for the Woodlands and Laboratory*, that was to be published by Cambridge University Press.

Many of Harry's books had been published by Macmillan and as Selina's financial problems mounted her plight somehow came to the attention of the head of that publishing house, Frederick MacMillan. He kindly agreed to try to muster help. On 11 July 1913 he wrote to Bower, and several other old friends and associates of Harry,

> "I have undertaken to prepare a memorial to the Prime Minister asking him to recommend Mrs Ward, the widow of the late Professor Marshall Ward, for a Civil List Pension. Mrs Ward has been left with very narrow means, and in view of her husband's valuable scientific work there is every reason to think that such a memorial, if signed by a few eminent botanists, would be favorably considered.
>
> I write to ask whether you will kindly append your name to the memorial… ."[03]

On 14 August, the letter signed by Thiselton-Dyer, Prain, Darwin, Scott, Seward, Bower, Bayley-Balfour, Vines and the Master of Christ's College was sent to the Prime Minister, Mr HH Asquith.[04]

Civil List Pensions were awarded to "those who by their personal services to the Crown, by the performance of duties to the public, or by their useful discoveries in science, and attainments in literature and the arts, have merited the gracious consideration of their Sovereign and the gratitude of their country". In deciding whether to award pensions to scientists or their dependants, it was Asquith's practice seek the advice of the Royal Society. In this instance its five-man Scientific Relief Committee meeting on 16 October 1913, and including two of Harry's old friends, Sir David Prain and Professor EB Poulton, decided, "in the case of Mrs Marshall Ward ... to recommend strongly the application to the favorable consideration of the Prime Minister on the grounds of the late Professor Marshall Ward's eminent services to botanical science, *both pure and applied* [author's italics]".[05]

A 'strong recommendation' was the highest category of approval. The same Committee on 16 October 1913 agreed only 'cordially to recommend' a second widow's case and refused to approve a third widow's case arguing that her husband's distinction as a scientist was not sufficient to justify support.

Whitaker's Almanak for 1915 records that an annual pension was granted to Selina. However, it was relatively small, only £40, in spite of the relative merit of her husband's achievements. The inference is that her personal circumstances were not as severe, and Harry's death was not as premature, as those of some of the other supplicants. All the evidence indicates that Selina's financial position continued to worsen. On 5 April 1914, Harry's father, Francis, died at the age of 83, but he left no inheritance to help his daughter-in-law and granddaughter. (Indeed, the old man, who lived with his daughter Nellie, had latterly been in receipt of a small pension from his son's estate and had indirectly been an additional burden upon Selina.) In Lyte's biography of Frank Kingdon-Ward,[06] he records that by 1920 Selina and Winifred were in straightened circumstances and forced to leave Cambridge for a small hotel in Bayswater, West London. Only two years later, on 18 June 1922, Selina died after a long illness.

Financial problems thus dogged the life of Harry's family after his death, just as they had dogged his own early life. Despite his exceptional talents, committed so selflessly to the good of future generations, whatever wealth and possessions he had accumulated in his lifetime were soon gone.

Harry might have left his family poor but to the plant sciences he left a rich legacy. He was a leader among the new breed of professional, academic scientists whose duties or activities were divided almost equally between research, teaching and administration; he managed to excel in each branch. The first branch, research, is the glamorous component for, apart from its intrinsic excitement, it can bring fame – if not fortune. It is the easiest to judge since reports are written to inform current and future generations. The record stands for all time. Teaching and, some, administration can be of at least as much worth, but their value is more subtle and the individual's role is harder to trace. Let us take each, in turn, and then attempt some synthesis.

In his research work Harry had many resounding successes, starting with his very practical demonstration that plant diseases, such as those caused by rust fungi, could be transmitted between plants by spores carried on the wind. Equally remarkable, however, but usually overlooked, was the strength and clarity of his argument that disease levels are affected by both the environment and management strategies. The implied lesson, that farmers could take simple measures that would do much to reduce the spread of infections, was slowly learned, helping to change notoriously conservative farming practices. Thus, by paying attention to spatial patterns of planting and natural vegetation, farmers found they could make it physically more difficult for spores to be carried from one plant to another: by making a strategic choice of the species and cultivars to grow, they found they could reduce the chances of spores landing upon genetically compatible plants.

In other circumstances, he might have continued to refine his epidemiology and field pathology but, beginning at Owens College, his opportunities lay more in laboratory-based work, where he could combine pathology with his interest in physiology. The most obvious outcome was his pioneering work on the enzymology of the plant-pathogen interaction (chapter 5), for which he has received well deserved recognition. What is much less well known, even among physiological plant pathologists, is that starting with studies of the rust, *Puccinia dispersa*, on brome grasses, which were published in 1902, he was the first to *describe* what Stakman[07] subsequently christened the hypersensitive reaction (HR) and he was the first to *recognise* the fundamental importance of the response among those plant reactions that resist infection.[08] We know now that HR, with its characteristic rapid host-cell death and subsequent necrosis, constitutes one of the primary mechanisms of resistance to plant disease.

Harry's observations of rust-infected grasses, and more critically his interpretation of them, are laid out more fully in the last paper he ever wrote, 'Recent researches on the parasitism of fungi' (1905).[09] With the help of Mr Biffen (of whom, more later) who had bred lines of wheat susceptible or immune, i.e. resistant, to infection by the rust fungus, *Puccinia glumarum*, Harry had continued studying the development of a fungus as it attempted to establish a parasitic relationship with a potential host plant. He extended his observations made on bromes by finding that, while stomatal penetration occurred at similar rates on both susceptible and resistant wheats, hyphae in the latter began to shrivel and lose vitality four to six days after inoculation, as they attempted to penetrate host cells. Most importantly, host cells surrounding those hyphae also began to degenerate, with loss of both their nuclei and chloroplasts. The phenomenon was localised to a 'nest' of cells around the invading hypha or hyphae. He concluded that, through their over-vigorous attack, the hyphae had killed the host cell(s) and so starved themselves. The modern interpretation is not dissimilar. Challenged cells of the resistant host are able to recognise the earliest signs of attack and undergo 'hypersensitive' cell death: physical and chemical changes convert what would have been a source of nourishment for the fungus into a zone that is not only nutritionally inaccessible but which is positively inhospitable to it, being packed with highly reactive and toxic free radicals and phenolic compounds. The significance of the HR, or cell-suicide, is that it underpins the effective response of most crop plants to their biotrophic fungal pathogens (and to viruses). Moreover, as 20th century plant breeders attempted to produce new, disease-resistant cultivars they focussed their attention, unwittingly at first, on genes controlling the recognition mechanisms which trigger the HR.

> Perceptively, in 1902 Harry had written[10], "All the evidence points to the existence, in the cells of the fungus, of enzymes or toxins, or both, and in the cells of the host-plant of anti-toxins or similar substances, as the decisive factors in infection or immunity, although I have as yet failed to isolate any such bodies".

Unfortunately, he made little headway in the last years of his life, not so much because he was distracted from his research by pressures elsewhere, but because he went down the wrong track. He decided to switch the focus of his studies of maceration from *Botrytis*, the fungus he had used in his earlier days at Cooper's Hill (chapter 5), to two of the rust fungi which were currently pre-occupying him, *Puccinia dispersa* (*recondita*) and *P. glumarum*. He switched from a necrotroph which, with hindsight, might be expected to secrete detectable amounts of the enzymes for which he was looking, to two biotrophs which, as we now know, are extremely conservative, secreting enzymes only in small amounts and in a pattern closely localised around their hyphae. Thus, by focussing on these rust fungi, he was unknowingly reducing his chances of detecting enzyme activity. Nevertheless, by the time of his last paper in 1905, Harry was able to leave to the world the rudiments of knowledge from which his successors have been able to build our modern molecular explanation of the way plants recognise and resist microbial infection. His fundamental work on the way enzymes are the foremost weapon used by pathogens to attack plants, and the way in which a plant may defend itself by sacrificing 'hypersensitive' cells, are at the heart of our current understanding of infection and resistance mechanisms, and of plant breeding. He was truly a founding father of physiological plant pathology.

Harry's successes were not confined to plant pathology but the same novel perspective often characterised his approach to different subjects because he was guided by his recognition that microbes are affected by their environment. So, in his studies of the microbiology of drinking waters, he focussed on the potentially bacteriocidal properties of sunlight while addressing broader questions about water storage. His findings contributed to altered policies among the water companies for they were given the confidence to build huge, open reservoirs in regions of high rainfall often far away from the growing cities which they supplied. Harry's studies of ginger beer produced less tangible results but, nevertheless, were highly respected at the time for their quality and, generally, for the way in which they advanced contemporary views on symbiosis and the interactions that may occur between species in mixed microbial communities.

Another example of symbiosis to which Harry dedicated considerable effort, and which has not previously been mentioned here, was the relationship between root-infecting nitrogen-fixing bacteria and their legume hosts. His efforts in connection with the latter project should probably be judged as partly but not wholly successful and their story reveals more about Harry's relationship with Thiselton-Dyer. Dyer claimed[11] that he "put him [Ward] onto the Nitrogen problem", just as he claimed similar credit for putting Ward onto the Ginger Beer problem. When Harry began his studies in 1883 the general opinion was that the swellings or tubercles (today called nodules) on the roots of leguminous plants housed parasitic bacteria-like organisms called bacteroids. In the years 1886 to 1888, thanks to the endeavors of two Germans, Hellriegel and Willfarth, it was finally proved to the satisfaction of all that legumes could use atmospheric nitrogen, fixation of which occurred in the swellings on the legume's roots. It was not known however whether the nitrogen was fixed by the bacteroids, or by the host cells which they stimulated to be formed. Harry started by confirming for the broad bean, *Vicia faba*, the results of Hellriegel and Willfarth. He then went on in 1887 to show for the first time that a micro-organism (identified in 1888 by Beijerinck as *Rhizobium leguminosarum*) invaded the root via the root hairs. "A hypha-like infecting tube or filament, … passes down the cavity of [the root hair], traverses the cortex of the root from cell to cell, until its tip reaches the innermost cells of the cortex, where it branches and stimulates these cells to divide and form the young tubercle."[12] Once again, Harry's manipulative skills and eye for microscopic detail had paid rich dividends.

Where he failed in his continuing researches, switching his attention in vain from broad bean to peas, was in making any progress in placing the site of N-fixation in plant or bacteroid cells. Given the lack of biochemical or microscopical tools for investigating such a complex relationship, Dyer's criticism of Harry's progress in this particular area seems harsh. "He had a woolly side to his brain and in spite of all that I could do he never got hold of the stick by the right end. … When he was wrong … there was no getting him back again."[11] If there was a criticism to which Harry was susceptible it was that, once again, he spread his energies too widely. He did undoubtedly have a tendency to lose focus.

While there is overwhelming evidence that Harry's own research and writings influenced the growth of physiological plant pathology after his lifetime, the personal connections he left were few. Unlike his idol, De Bary, he did not have a string of postgraduate students and visiting workers passing through his laboratory who would, in their turn, make important discoveries and train another generation of leading scientists. Perhaps his best known research student was FT (Tom) Brooks. Completing his studies on *Botrytis cinerea* at the time of Harry's death, Brooks agreed to take over his considerable teaching load in plant pathology and mycology at very short notice. Brooks' career was not irreparably damaged by his generosity nor by his antipathy to AC Seward who succeeded Harry as Head of Department. (Apparently fuelled by the conviction that no one could measure up to his beloved Marshall Ward, Brooks' antipathy even led him to accuse Seward of being a German spy during the Great War.[13]) In spite of such idiosyncrasies, Brooks became a leading plant pathologist of his generation, not to mention Professor and Head of the Botany School (1936–48). But Brooks was a traditional plant pathologist, not a physiologist. He was most at home unravelling infection cycles and working out control methods. He had none of the genius or originality of his supervisor, he was largely unknown outside Britain and, except for being an outstanding teacher, has been largely forgotten even by his fellow countrymen. Others of Harry's Cambridge students, such as Dorothea CE Marryat, who took forward his studies of the histology of the immune response of wheat to rust fungi[14], made no lasting impact themselves on plant pathology.

Those who benefited most from Harry's scientific legacy were men like William Brown and RKS (Ronald) Wood who, as we have seen in chapter 5, had no direct links with him. It is difficult to understand why Harry left no direct inheritors. He was overburdened with responsibilities, all of which he took very seriously, and he might have been a difficult man where his staff were concerned, but at least one of his students found in him all the inspirational and personal qualities one would look for in a supervisor. The views of 'MD', published as a footnote to the 'Obituary' Dyer wrote in *New Phytologist*[01] are quoted in full.

> "Never shall I forget his kindly welcome when I came up to Cambridge to begin research under his supervision, at all times his attitude to his students was that of comrade and fellow-worker, rather than Professor and critic. Whilst always ready to help others, he was as ready to learn from anyone, even his most insignificant student. Even during term-time, when he was already over-busy with lectures and syndicates, he rarely allowed a day to pass without visiting each of his research students, and it was indeed remarkable how he would pass from one to another, and in a moment grasp fully the exact stage which each research had reached. His criticisms were often severe and unreserved, and his praise was very rare, but when any real success had been gained his few words of congratulation were so spontaneous and sincere, that they gave encouragement for many days to come.
>
> Everyone understood that Professor Ward would not tolerate slackness in any form; there must be no drones in the hive, but Staff, Students and Laboratory Assistants must all alike be keen workers. As a lecturer, he was unrivalled for clearness of speech, lucid explanations and the power of awakening and riveting interest: by vivid description, rapid drawings and the constant application of his knowledge to matters of practical importance, he kept his hearers' attention from beginning to end; and by impartial discussions of opposing theories, he cultivated in them independence of thought and a desire for further study of the subject in hand.

> Whether at home or in his laboratory Professor Ward was never too busy to give advice to any who sought his help, and he was continually on the watch for opportunities when he might use his influence on behalf of those he considered worthy. His untimely death has removed from our midst a great teacher and a true friend to those of us who knew him; but we shall never forget him nor cease to be grateful for the privilege of working with and for him. His students are now living in very many parts of the world, and through them, as well as through his public work and writings, his influence must be felt for long years to come."

Perhaps the last sentence leads to the answer. Perhaps he was too much ahead of his time. While he could find those who could learn and carry around the world the valuable lessons of traditional plant pathology that he was able to teach them, he was unable to find, or incapable of identifying, a student with a genius to match his own: a student who could bring together the complementary disciplines of pathology and physiology.

From this general observation there is one outstanding exception and that involves two of North America's most distinguished pathologists, Edward Monroe Freeman and Elvin Charles Stakman (chapter 7). The connection began with Freeman (Fig. 8:1) who sought out Harry because it was clear to

Fig. 8:1. EM Freeman, a leader of plant pathology at the University of Minnesota and through North America. (From the private collection of Paul Peterson, by permission.)

him that "Marshall Ward was a major prophet at the time, and his every pronouncement was regarded pretty much as the revealed truth".[15] Freeman was an exceptionally able botanist, yet in 1900 he was doing mundane work in his home state of Minnesota, recording and listing rust and powdery mildew diseases. At the start of the twentieth century most scientific information was still passing from east to west across the Atlantic and, to better himself, Freeman chose to spend the year 1901 in Cambridge. In Harry's laboratory he became infected with the germ of pathology.[16] He became involved in studies of rust on the grass *Lolium temulentum* and was instrumental in helping Harry to refute Eriksson's 'Mycoplasm Theory'. Luckily, or maybe wisely, he managed to avoid being tarnished as a supporter of the ill-fated 'Bridging Theory'.

Freeman's career blossomed on his return to the University of Minnesota. Building on the work he had done in Cambridge on rust fungi, he made important contributions to contemporary knowledge of the epidemiology of cereal rusts; he published *Minnesota Plant Diseases* (1905), one of the first textbooks of plant pathology in the USA, no doubt having been stimulated by Harry's earlier textbook *Disease in Plants* (1891); and it was thanks to Freeman that plant pathology was gradually established as an independent discipline within the University of Minnesota. (It is no coincidence that St Paul, MN, is today the home of the American Phytopathological Society.) The student whom Freeman taught, in turn, was Stakman, a man whose achievements are legendary. Stakman's long career was dedicated to defining variability in pathogens (see Box 8:1). His work enabled breeders to know their enemy, to direct their energies toward incorporating into plants the genes that gave resistance to the most prevalent and destructive races of the pathogen. Since history has shown those races are constantly evolving, responding to the resistance genes deployed by breeders, an on-going and seemingly never-ending arms race has resulted. But, at least, thanks to Stakman and others, we understand its rational basis.

If it is agreed that a Professor should be judged not only on the output of his own laboratory but on the output of his more junior colleagues, since he has a responsibility to guide, direct and encourage them, then posterity must surely judge Harry extremely successful. He might have had difficulty in finding research students who could match his own brilliance but he had no such difficulty in identifying brilliant colleagues. His most significant personal interactions, which led to botany being taken forward, were with Rowland Harry Biffen and Frederick Frost Blackman.

No one made more of a contribution to the rational basis of crop improvement than Biffen, who has often been called 'the father of plant breeding'. Biffen joined the Cambridge School of Agriculture shortly after its foundation in 1899, just in time for the rediscovery of Mendel's Laws of Inheritance in 1901. He realised immediately what they meant for the future of plant breeding which hitherto had had little scientific, rational direction. Speculating that physiological as well as morphological traits were inherited in Mendelian ratios, he gave careful thought to the physiological trait(s) that

Box 8:1. From Stakman to the release of Genetically Modified pathogens

The method of analysing physiologic races developed by Stakman can show the success or otherwise of control strategies, for example the eradication of barberry, the secondary host of stem rust, which allows it to reproduce sexually. As the result of a US-wide programme led by Stakman, the number of rust races declined from 30 in 1918 to only 4 in 1978.[17] Race analysis was widely and profitably employed for over half a century, with new levels of sophistication being introduced only when host types differing in one single gene were developed, first by orthodox breeding methods and later by genetic manipulation. These nearly genetically identical lines are capable of differentiating an almost infinite number of lines of the fungus.

The potential for *natural* variation within a species of a pathogen, which enables it to extend the list of cultivars that it can attack, and which is controlled by one or a very small number of genes, does not extend to enabling that same species of pathogen to change to attack new or different host species. Aggressiveness, or virulence (in the botanical sense), on a natural host might change, but the evidence from a century of plant pathology is that pathogens do not jump from one host species to another. The stability of species-species relationships has been widely accepted since Stakman's day until very recently when, despite any firm evidence to the contrary, the possible release of genetically modified pathogens to control weeds or drug-producing crops, such as coca, has initiated new worries about whether or not pathogens may change hosts.

might be most amenable to proving his point. He had always had a keen interest in fungi and now, with advice and encouragement from the distinguished pathologist Harry Marshall Ward, he picked for his studies the resistance of wheat to yellow rust (*Puccinia striiformis*). The experiments proved Biffen's point; he crossed the resistant cultivar 'Rivet' with the susceptible cultivar 'Red King' and found that in the second (F_2) generation the ratio of susceptible to resistant progeny was very close to the 3:1 ratio which Mendel's Laws predicted, if the character was controlled by a single gene, that is if it is monogenic. Biffen's critics pointed out that there was little point in breeding new cultivars if the Bridging Theory was correct and pathogens could change host, but Biffen's announcement (1907)[18] stimulated other researchers to study the inheritance of resistance in different plant-pathogen systems. The trickle of research became a flood after Stakman destroyed the Bridging Theory and by 1934 over 200 papers had been published with resistance to 23 diseases in 13 different crops being reported as being monogenic.[19] Biffen's reputation and influence grew and grew, not least because he bred two commercially very successful wheat cultivars, 'Little Joss' and 'Yeoman'. By 1912 he had been made the first Director of the Plant Breeding Institute in Cambridge, a body which for over 50 years was to be among the world's leaders. Later he was instrumental in the establishment of the National Institute of Agricultural Botany, also in Cambridge. Despite the differences in their ages, Ward and Biffen had great respect for each other, recognising the complementarity of their interests. We can only speculate what great things they might have achieved together had not Harry's life ended so prematurely.

Blackman was another young man whose early career was helped by Harry and who was in later life to achieve considerable distinction – this time in plant physiology. Apart from promoting Blackman from Demonstrator to Lecturer in 1897 and offering constructive interest in his work, Harry was able to give practical help, using his influence within the British Association to persuade that body to make small grants to Blackman for the purchase of apparatus. A small committee was set up by the BA to monitor how the money was spent in Blackman's investigations of 'Respiration in Plants'. He was in fact studying photosynthesis as well as respiration and it was his paper, published in 1905, on the factors limiting photosynthesis which made his name. Blackman and Mathgel grew their plants under different combinations of environmental conditions, controlling the concentration of carbon dioxide in the atmosphere, light intensity and temperature. In turn, two of the parameters would be held constant while the third was varied. They found that the rate of photosynthesis, i.e. fixation of carbon dioxide, was dependent on temperature when the light was strong and amounts of carbon dioxide were at normal atmospheric concentrations or lower. This suggested that, when light is not the limiting factor, carbon dioxide fixation is based on normal, temperature-dependent biochemical reactions. In contrast, with the combination of higher concentrations of carbon dioxide and little light there was no temperature-dependence. This suggests that, when light is limiting, the light-induced reactions are independent of the temperature. Blackman and Mathgel's work thus began to reveal what we regard today as a fundamental truth about photosynthesis, it is comprised of both light and dark reactions which are, respectively, independent of and dependent upon the activity of enzymes.

Blackman's name has often been conjoined, in the Liebig-Blackman Law, with that of the famous German agricultural chemist, Justus Liebig, since the latter proposed his closely comparable 'law of the minimum' in 1841. In relation to the effects of the availability of mineral nutrients on plant growth, what Liebig's law says is that growth of the plant is dependent upon the amount of that nutrient presented to it in minimum quantities. For example, growth (or photosynthesis) will be limited by low phosphorus (or temperature) even though the plant has abundant nitrogen (or high light). When applied to physiological ecology, the law explains why a limiting environmental factor often determines the geographic range of a plant species.

In the view of one of his contemporaries, J Reynolds Green, who was a Fellow and Lecturer at Downing College from 1902 to 1907 and who published in 1914 his lengthy *History of Botany in the United Kingdom*,

"[Ward] built up a reputation for himself that was almost unique ... His research at the borders of health and disease (physiological pathology) was in some ways the most remarkable of the century".[20] Not only this, Green opined, Harry was "also a great figure in education and administration". Other commentators agree but, concerning Cambridge, one dissenting voice is notable and, not surprisingly, it is that of Thiselton-Dyer. He wrote, "... although the Cambridge men all speak of him with respect and affection, they all agree that as Professor he was not a success. He kept his staff at arm's length and never appeared to have an intelligible policy they could follow."[11] The specific antipathy that Walter Gardiner held towards Harry has been described (p. 120), and exactly what Dyer was referring in the end of this last remark is not known, but such an opinion seems especially difficult to justify given both Harry's success in obtaining a major new building for botany in Cambridge and the revision of teaching which he undertook in the spirit of the 'The Cause'.

In his lectures he may not have had the youthful enthusiasm that so impressed his students at Owens College, and he probably overloaded himself, but he still had the capacity to inspire young men. John Ramsbottom, who became Keeper of the Herbarium at the British Museum (Natural History), one of Britain's greatest mycologists, was also a student at Cambridge and witnessed Harry's teaching at first hand. He recalled, "I received my first botanical lectures from Marshall Ward, and well remember the thrill of his philosophical comments".[21]

Harry's contributions to the administration of national and international bodies began long before his Cambridge days but they grew and grew, giving him no respite in his final years – even if he had sought it. Through his Fellowships of the Linnean Society and the Royal Society he was asked to serve on numerous committees, large and small. Similarly through the British Association, its meetings and its works, he promoted and influenced the development of botanical sciences. One of the institutions whose creation he strongly encouraged was the *Annals of Botany* (chapter 5) and he supplied the journal with a steady stream of his papers, not because it seemed an easy route to publication but because the journal enshrined the New Botany about which he was so enthusiastic. Another institution which he helped to found, and which has not been mentioned so far, is the British Mycological Society (BMS).

The part that Harry played in the early development of the BMS neatly exemplifies his wider contributions to the history of botany. The origins of the BMS, second only in age to the Société Mycologique de France founded in 1884, lie with a succession of highly talented amateur mycologists who met regularly to spend their days collecting fungi from the field and their evenings eating and drinking. Their focus from 1868 was the Woolhope Naturalists' Field Club, based near Hereford, but by the 1890s the centre of their activities had moved to the Yorkshire Naturalists' Trust. Recognising the wider national interest in fungi, its leading members decided to found the BMS whose first meeting was held, appropriately, at Selby, Yorkshire

(1896).[22] Although the first President, George Massee, could be classed as a professional, since he was employed at Kew (see p. 33), its other officers were not. The Secretary, Carleton Rea, was by profession a barrister, while the Treasurer, Charles Crossland, was a butcher. Nearly all of the 47 members in the first published list were amateur mycologists, gentlemen whose interests in fungi were limited to collecting and describing.

To have Harry Marshall Ward FRS as a founder member was something of a catch for the young Society; he was a distinguished academic with an impeccable address. In his inaugural Presidential speech Massee had said, "Now that Professor Marshall Ward is located at Cambridge, the growth of the British School of Mycology is simply a matter of time". Time was not, of course, on the side of either Harry or the BMS, but the Society's survival and success in those early years was due in no small part to his efforts during two years as President (1900–01) and five years as Vice-President (1902–06).[23] He attended forays all around Britain, though it was noted that like several other eminent mycologists 'he had no eye for species', and he persuaded members to contribute to a 'Printing Fund' that paid for illustrative plates to be included where appropriate in papers in the Society's *Transactions*. He was clearly determined that no author should suffer as he had done at the hands of the Royal Society (p. 108). Biffen was the next President (1905) to come from a background of experimental science and it was not until the 1920s that the 'professionals' came to dominate the Society, but the early election of Harry to the Presidentship signalled that the Society would be the natural home for the growing numbers of professional mycologists who would follow. The BMS was the 'home' of British plant pathologists until 1981 when the British Society for Plant Pathology was founded.

The final part of Harry's legacy is comprised of the books and popular articles he wrote, or translated. In some ways they can be seen as the tangible products of his mission to promote the New Botany. Thus, in translating the Sachs's lectures (1887) or Hartig's book on tree diseases (1894), he brought German knowledge to an English-speaking audience. It was not in his nature to worry that Hartig's book might compete with his own, *Timber and Some of Its Diseases* (1889), written only five years earlier largely for the benefit of his students at Cooper's Hill. Viewing his books in retrospect, there is a suspicion that later in his life he came to appreciate the importance of the older botanical traditions to agriculture and forestry. Thus, in his Preface to *Trees: A Handbook of Forest Botany*, he wrote, "Considerable experience has convinced me that the neglect of the older methods of observation of the living plant, which rendered Botany so exhilarating to the naturalist in pre-laboratory days, is to a large extent our own fault. The textbooks of today, excellent as they are in most respects, are apt to lack that peculiar kind of stimulus, so necessary for young students, which drives the reader to observe and think for himself, and not through the eyes – or worse still the words- of others." His book on *Grasses: A Handbook for Use in Field and Laboratory* (1901) was similarly written for the benefit of students. He observed, "In view of the importance of our British grasses in agriculture, I have here put together some results of observation and reading in the hope that they may aid students in recognising easily our agricultural and wild grasses". The little book (190 pp. plus many original illustrations) still provides today an easy and accurate route to the identification of the commoner British grasses.

Harry's achievements and endeavours in teaching many aspects of botany would be of only of limited interest if it were not for the two books he published on the principles of plant pathology, *Diseases of Plants* (1891) and *Disease in Plants* (1901). Their importance lies in the fact that before them there were very few books written on the subject and the authors of some of those books were unreliable in the opinion of many commentators.

Let us take the unprejudiced view of a famous North American bacteriologist, Erwin Smith, reviewing *Fifty Years of Pathology* – from the perspective of 1926.[24] He recalled that in the years up to 1887, there were in English

> "few books of any sort, good or bad. I recall Berkeley's *Outlines* [of British Fungology] with a chapter on plant diseases which was good but old (1869) and very incomplete, and Worthington G Smith's *Diseases of Field and Garden Crops*, newer (1884) and more extensive but less dependable. There was also a little book on *Rusts, Smuts, Mildews, and Moulds*, by M C Cooke (1865), which did service of a sort in default of something really good. I used also a little German book by George Winter (1878). We had to be content with books and papers in Latin, French, and German, and largely with systematic treatises on fungi. So we read Berkeley and Tulasne, Corda and Fries, Montagne and De Bary, and all the older writers, and English and American plant pathology in this decade was little more than a sublimated mycology."

To this list, in the categories 'old' and 'in German', Erwin Smith might have added one more solid tome, Julius Kühn's 1858 work, *Die Krankheiten der Kulturgewachse, ihre Ursachen und ihre Verhutung*, or *The Diseases of Cultivated Plants, Their Cause, and Their Prevention.*

Even after 1887 things improved but little for, as typified by the first book of pathology published in the USA, *Fungus Diseases of the Grape and Other Plants and Their Treatment* (1890) by Scribner, new books typically concentrated on one disease or one crop. And what, more importantly, of the quality these authors and their books?

For all his brilliance, Berkeley was not an experimentalist; nor were Kühn (a mixture of scientist and farmer), Cooke, or Worthington-Smith. Writing on the 50th anniversary of the BMS, Ramsbottom[23], then President, commented that Cooke had a wide knowledge of fungi but little critical faculty. To his dying day, Cooke denied that rust fungi could include in their life cycle two unrelated host plants, and also that lichens combined fungi and algae in their bodies. Worthington-Smith's credentials were those of an architect, turned botanical illustrator. Sadly, the artistic quality of his drawings was not always matched by the accuracy of his observations. He is memorable also for an enormous scientific *faux pas*, in 1875 he announced that he had found the resting spores of the potato blight fungus which had previously eluded, among others, Anton de Bary. Such a conclusion was totally unjustified, there may have been a few resting spores in his microscope preparations but they were amongst an amateurish mess comprised of the bodies of several fungi.[25,26] Erwin Smith did not mention Massee, – who published in 1891 *British Fungi, Phycomycetes and Ustilagineae* [the latter are Smut fungi, important pathogens infecting plant reproductive organs]. Again, we turn to Ramsbottom[23] for comment: "... a most unsatisfactory production ... Massee had as wide a knowledge of mycology as anyone of his period, but was quite unreliable. ... one cannot accept his original statements on trust. All must be verified. They may be brilliant interpretations or they may be absurdities simulating boyish pranks in a superfluity of naughtiness".

Harry Marshall Ward's books were much, much better than any which had gone before, or which were available at the time. They were better because they were authoritative and accurate; they drew upon first-hand knowledge gained through countless hours in the laboratory. They were the work of a professional.

Through his research, and teaching, both undergraduate and postgraduate, promotion of journals and societies, and through his own books, no man did more than Harry Marshall Ward to promote 'The Cause' – to bring the New Botany to Britain. But how much was he 'his own man'? A theme recurring throughout this book has been Thiselton-Dyer's relationship with Harry. Was Dyer the puppet-master pulling the strings? Did he exert as much influence over Ward as he would have us believe, or did he just want to bask in any glory that reflected from Harry's achievements? The time has come to attempt a final answer to such questions.

There is no doubt that Harry was uncertain of himself, at least in his younger days. Letters from Ceylon show this, as does a letter he wrote to Balfour soon after his arrival in Manchester.[27] (Remember he and Balfour had shared a tiny cabin for several weeks.)

> "My Dear Balfour,
>
> If I may address you in consideration of the friendship we formed on board the <u>Ava</u>. You have become such a swell since you returned from Socotra, that I almost expect you have forgotten me and my journey to Ceylon: I have not forgotten you, however, and hope that now I am once more in England, I may hear from you now and again, if your very busy life admits of correspondence."

Dyer suffered no such self-doubts: all who came into contact with him would agree that he was undoubtedly bossy and manipulative, and sometimes he would put the interests of institutions before those of individuals. However, Harry was his brightest student, his protégé, and, perhaps because of Harry's occasional need of self-confidence, they settled into a father-and-son-like relationship. Dyer even wrote to Bayley-Balfour, "Ward certainly never repudiated his parental relationship with myself. He probably assumed it to be a matter of common knowledge". In the same letter Dyer[11] betrays the anxieties and pride of a father,

> "... if he had not come into my hands he would have probably ended his days as a master in a Board School. For, having started him in Botany, I had practically to find him the means of livelihood afterwards and push him along... He never had much confidence in himself and was a little despondent. From 1879–89 it was a constant anxiety to me how to keep him going. He always seemed to require ... stimulus. But I never came across a man who responded to it more admirably. In all these researches [Cooper's Hill, Lily Disease and Ginger Beer] he showed a quality which, I think, amounts to positive genius, and which I have not myself come across, at least to the same extent, in anyone but Mr [Charles] Darwin himself".[01]

Harry reciprocated with a son's wish to please his father. It was always to Dyer that he reported his successes, as well as his failures. On learning that he was to preside over Section K of the BA meeting in Toronto he wrote to Dyer,

> "Now, to crown all, comes your kind letter [of congratulation]. I may truly say that I value the unstinted approbation you accord me more highly than any I have ever received; and you should not forget that anything I may have been able to do has been very largely due to your stimulating influence on me from the first steps of my career. I cannot sufficiently thank you for all your goodness in rendering my work possible, and for many an encouraging word at just the right dull moment."[28]

His need for Dyer's good opinion is seen again in the letter he wrote to Dyer after the latter had congratulated him in November 1895 on the award of the Cambridge chair: "Of all the many kind letters of congratulations I have had, none has given me the sharp feeling of pleasure that yours has".[29]

Dyer was a man of considerable achievements and influence, and through his long and varied career he came to know personally all the great British botanists and most of great European and American botanists of the age. His opinion that Ward ranked second only to Darwin must therefore be taken seriously. But, recognising possible personal bias, we should conclude with the comments of two other, more detached, contemporary observers, Sydney Vines and Frederick Bower. In a very few words they capture Harry's strengths and his weaknesses, both as a man and as a scientist.

First, Vines:

> "I remember Ward as a genial companion, a man of varied interests, delighting especially in music; but the dominant impression is that of his whole-hearted devotion to his science; all else counted with him as nothing in comparison with that. No doubt this led him to impose too severe a strain upon a constitution never very robust; but such as he was, it could not have been otherwise".[30]

And now Bower:

> "Before all things Ward was an observer: in this respect day and night were to him alike, and he followed living organisms the clock round with a tenacity which can still be traced between the lines of his papers, and be appreciated even by those who never saw him personally at work. For him it was always the

problem of the living, not the dead organism; and inevitably his interest was deeper in ontogeny than phylogeny [the development of an organism rather than its systematic relationships (see chapter 1)]. His close observation led to a certain defect of style. His papers are apt to read rather like laboratory notes, than as the outcome of literary skill; in fact their prime merit became a blemish."[31]

Harry is buried, alongside Selina, in the small, hidden cemetery of All Soul's Chapel in the Parish of St Giles with St Peter, Huntingdon Road, Cambridge. The large, simple and erect headstone bears clear lettering in bold relief. It records his two chairs and his two fellowships. On it is inscribed his epitaph: 'Laborare est orare', – work is prayer.

Harry Marshall Ward's world was his science and, through his work, he improved that world. No man can do more.

References

Sources of unpublished, archival material are abbreviated as follows:

BL, British Library, London (MacMillan letterbooks)
BO, Bodleian Library, Oxford University (Archives of the British Association for the Advancement of Science)
GU, Glasgow University
HU, Harvard University, Farlow Library
NHM, Natural History Museum, London
NNS, Nottinghamshire Naturalists' Society
OUM, Oxford University Museum
RBGE, Royal Botanic Garden, Edinburgh
RBGK, Royal Botanic Gardens, Kew
RS, Royal Society of London

Chapter one: To Ceylon for coffee

01 RBGK, family letters, 20 April 1881.
02 Mackenzie, J.M. 1990. Imperialism and the Natural World. Manchester University Press, Manchester.
03 Mackenzie, J.M. 1991. Plant collecting and imperialism. Pages 8–14 in: Reginald Farrer. J. Illingworth and Jane Routh, eds. Center For North West Regional Studies, Lancaster University.
04 Ludowyk, E.F.C. 1966. The Modern History of Ceylon. Weidenfield and Nicholson, London.
05 Mills, L.A. 1964. Ceylon Under British Rule 1795–1932. Frank Cass, London.
06 Anonymous. 1880b. Coffee Leaf Disease. Second report by H Marshall Ward. Colombo Sessional Papers **L**, 555–566.
07 Gregory, P.H. 1961. The Microbiology of the Atmosphere. Leonard Hill, London.
08 Schneiderhan, F.J. 1933. Phytopathological Classics Number 3: The Discovery of Bordeaux Mixture. American Phytopathological Society, St Paul, Minnesota.
09 Anonymous. 1881. Coffee Leaf Disease. Third report by H Marshall Ward. Colombo Sessional Papers **XVII**, 227–262.
10 RBGK, family letters, 13 January 1880 (dated in error 13 January 1879).
11 RBGK, family letters, dated 18 January 1880.
12 RBGK, family letters, dated 10 August 1880.
13 RBGK, family letters, dated 30 January 1881.
14 Thiselton-Dyer, W. 1901. Henry Trimen 1843–1886. Obituary Notices. Proceedings of the Royal Society of London 75, 161–165.

15 RBGK, family letters, dated June 1880.
16 Anonymous. 1880a. Coffee Leaf Disease. Preliminary report by the Government Cryptogamist. Colombo Sessional Papers **XVII**, 141–144.
17 NHM, dated 5 February 1882.
18 RBGK, family letters, dated 20 April 1881.
19 Thiselton-Dyer, W. 1911. Harry Marshall Ward (1854–1886). Proceedings of the Royal Society of London, B, 83, 1–14. (Re-published in 1913 in Makers of British Botany (ed. F.W. Oliver), pp.261–280. Cambridge University Press, Cambridge).
20 RBGK, Director's correspondence, 105, 6–7.

Chapter two: Becoming a botanist

01 RBGK, family letters, dated 15 January 1881.
02 Layton, D. 1993. Reading science: images of science in some nineteenth century reading lesson books. Paradigm 10 (April).
03 Russell, C.A. 1996. Edward Frankland. Chemistry, Controversy and Conspiracy in Victorian England. Cambridge University Press, Cambridge.
04 Haines, G. 1958. German influence upon scientific instruction in England, 1867–1887. Victorian Studies, March, 215–244.
05 Yeo, R. 1993. Defining Science . William Whewell, Natural Knowledge, and Public Debate in Early Victorian Britain. Cambridge University Press, Cambridge.
06 Layton, D. 1973. Science for the People. The Origins of the School Science Curriculum in England. George Allen & Unwin, London.
07 Argles, M. 1964. South Kensington to Robbins. Longmans, London.
08 Whitrow, G.J. 1972. Centenary of the Huxley Building: Personalities and Events in the History of Imperial College. (Pamphlet, Imperial College library)
09 Desmond, R. 1995. The History of the Royal Botanic Gardens, Kew. The Harvill Press, London.
10 Schmid, R. 1999. Bamboozled by botany, Beatrix bypasses bigoted biology, begins babying bountiful bunnies. Taxon 48, 438–443.
11 English, M.P. 2001. Leila Cooke's notes on the life of her father, M.C. Cooke. Mycologist 15, 91–93.
12 Thiselton-Dyer, W. 1911. Harry Marshall Ward (1854–1906). Proceedings of the Royal Society of London, B, 83, 1–14.
13 Thiselton-Dyer W. 1925. Plant biology in the 'Seventies'. Nature 115, 709–712.
14 NNS, Annual Reports and Transactions.
15 Williams, B. 1976. The Making of Manchester Jewry 1740–1875. Manchester University Press, Manchester.
16 Thompson, J. 1886. The Owens College: its Foundation and Growth. J.E. Cornish, Manchester.
17 Gillispie, C.C. 1971. William Kingdon Clifford. In Dictionary of Scientific Biography, III, 322. Scribners, New York.
18 Anonymous. 1876. The Times, Tuesday 26 December.
19 Bower, F.O. 1938. Sixty Years of Botany in Britain (1875–1935). MacMillan, London.

Chapter three: German lessons

01 Horsfall, J.G., and Cowling, E.B. 1977. The sociology of plant pathology. Pages 11–33 in: Plant Pathology: An Advanced Treatise, Vol. 1, How Disease is Managed. J.G. Horsfall and E.B. Cowling, eds. Academic Press, New York.
02 Large, E.C. 1940. Advance of the Fungi. Jonathan Cape, London.
03 Sparrow, F. 1978. Professor Anton de Bary. Mycologia 70, 222–52.
04 Drews, G. 2000. Anton de Bary, ein bedeutender Biologie, lehrte in Freiburg, Halle und Strassburg. Freiburger Universitatsblatter 149, 5–25.
05 Ward, H.M. 1888. Anton De Bary. Nature, Jan 26th, 297–299.
06 De Bary, A,H, 1863. Recherches sur le développement de quelques Champignons parasites. Annales Sciences Naturel (Botanique) 20, 5–148.
07 De Bary, A.H. 1876. Researches into the nature of the potato-fungus – *Phytophthora infestans*. Journal of the Royal Agricultural Society of England 12, 239–269.
08 Geison, G.L. 1995. The Private Science of Louis Pasteur. Princeton University Press, Princeton, NJ, USA.
09 Howard, M. 1961. The Franco-Prussian War: The German Invasion of France 1870–1871. Routledge, London.
10 Harris, H.F. 1945. Correspondence of W.G. Farlow. Farlowia 2, 36–37.
11 Jost, L. 1930. Zum Hundersten Geburstag Anton De Barys. Zeitschrift fur Botanik 24, 1–74.
12 Hill, T.G. 1931. The Advancement of Botany. Presidential Address to Section K. British Association for the Advancement of Science xx, 2–20.
13 Bower, F.O. 1925. English and German botany in the middle and towards the end of the last century. New Phytologist 24, 129–137.
14 Reess, M. 1888. Anton de Bary. Ein Nachruf. Berichte der Deutschen Botanicher Geselschaft 6, 8–26.

Chapter four: From industrial Manchester to leafy Englefield Green

01 Williamson, W.C. 1896. Reminiscences of a Yorkshire Naturalist. George Redway, London.
02 RBGK, family letters, dated 30 June 1881.
03 Pers. Comm., D. Brady, John Rylands Library, Manchester University.
04 Thompson, J. 1886. The Owens College: Its Foundations and Growth. JE Cornish, Manchester.
05 Kargon, R.H. 1977. Science in Victorian Manchester: Enterprise and Expertise. Manchester University Press, Manchester.
06 Ward, H.M. 1886. On the sexuality of fungi. (Reprinted from the Quarterly Journal of Microscopical Science, 1884.) Biological Studies of the Owens College 1, 23–63.
07 Ward, H.M. 1886. The morphology and physiology of an aquatic myxomycete. Biological Studies of the Owens College 1, 64–86.
08 Ward, H.M. 1883. On the coffee leaf disease of Ceylon. Manchester Literary and Philosophical Society Proceedings 22, 21–25.
09 Ward, H.M. 1884. On the fungus of the Salmon Disease – *Saprolegnia ferax*. Manchester Literary and Philosophical Society Proceedings 23, 29–32.
10 Tooley, O., pers. comm. (HMW's letter dated 10 September 1877.)

11 RBGK, family letters, dated 3 October 1880.
12 RBGK, family letters, dated 10 August 1880.
13 RBGK, family letters, dated 16 September 1881.
14 RBGE.
15 RBGE.
16 GU archive 488.
17 Boney, A.D. 1985. Appointment to a 'Crown' chair of botany: Glasgow 1885. The Linnean 1, 19–26.
18 Boney, A.D. 1998. The establishment of an Institute of Botany, 1879–1901. The Linnean 13, 15–37.
19 Nelson, E.C. 1990. William Ramsey McNab's herbarium in the National Botanic Gardens, Glasnevein (DNB). 1 Its early history and acquisition. Glasra 1–7.
20 Nelson, E.C. and McCracken, E.M. 1987. The Brightest Jewel. A History of the National Botanic Gardens Glasnevin, Dublin. Boethius Press, Kilkenny.
21 Bower, F.O. 1938. Sixty years of Botany in Britain (1875–1935). MacMillan, London.
22 GU archive 9976.
23 GU archive 9979.
24 Anker, P. 2001. Imperial Ecology. Harvard University Press, Cambridge MS.
25 Kingdon Ward, W. (unpublished). The Flower Chief. Archives of the Royal Botanic Garden, Kew.
26 Lyte, C. 1989. Frank Kingdon-Ward. The Last of the Great Plant Hunters. John Murray, London.

Chapter five: Work will enable us to forget

01 Kingdon-Ward, W. (unpublished). The Flower Chief. Archives of the Royal Botanic Garden, Kew.
02 Thiselton-Dyer, W. 1911. Harry Marshall Ward (1854–1906). Proceedings of the Royal Society of London, B, 83, 1–14. This paper was re-published in 1913 in Makers of British Botany (ed. F.W. Oliver), pp. 261–280. Cambridge University Press, Cambridge.
03 Williamson, W.C. 1896. Reminiscences of a Yorkshire Naturalist. George Redway, London.
04 Wilson, K. 1978. The origins of the Annals of Botany. Annals of Botany 42, 741–745.
05 HU, archive 48.
06 HU, archive 49.
07 RBGK, Director's correspondence 105, 35.
08 Ward, H.M. 1888. A lily-disease. Annals of Botany 2, 319–378.
09 Sumner, J.B., and Somers, G.F. 1947. Chemistry and Methods of Enzymes. Academic Press, New York.
10 De Bary, A.H. 1886. Ueber einige Sclerotinien und Sclerotienkrankheiten, Botanishes Zeitschrift 44, 377–474.
11 Jones, L.R. 1910. Bacterial soft rots of certain vegetables. II. Pectinase the cytolytic enzyme produced by *Bacillus carotovorus* and certain other soft rot organisms. Vermont Agricultural Experiment Station Bulletin 147, 281–360.
12 Garrett, S.D. 1975. William Brown 1888–1975. Biographical Memoirs of Fellows of the Royal Society 21, 155–174.

13 Brown, W. 1965. Toxins and cell-wall dissolving enzymes in relation to plant disease. Annual Review of Phytopathology 3, 1–18.
14 RBGK, Director's correspondence 52,45.
15 RBGK, Director's correspondence 55, 48.
16 Bower, F.O. 1938. Sixty years of Botany in Britain (1875–1935). MacMillan, London.
17 Bower, F.O. 1906. Harry Marshall Ward, FRS. Journal of Botany 44, 422–425.
18 RBGK, family letters, dated 3 October 1880.
19 OUM, correspondence of EB Poulton.

Chapter six: Ginger beer and a loss of focus?

01 Ward, H.M. 1898. Some brewing botanical problems. Journal of the Federated Institutes of Brewing 4, 334–363.
02 Ward, H.M. 1890. The fungi of fermenting vessels, vats, &c. The Brewers' Journal, 33–36 (continued) 116–118, 223–226, 408–412, 524–527, 623–627, also (1891) 40–42.
03 RBGE, letter from Thiselton-Dyer to I. Bayley-Balfour, dated 30 December1907.
04 English, M.P. 1987. Mordechai Cubitt Cooke. Biopress Ltd., Bristol.
05 Ward, H.M. 1892. V. The Ginger-Beer plant, and the organisms composing it: a contribution to the study of fermentation – yeasts and bacteria. Philosophical Transactions of the Royal Society, Series B, 183, 125–197.
06 Ward, H.M. 1897. On the Ginger Beer plant. Annals of Botany 11, 341–344.
07 Russell, C.A. 1996. Edward Frankland. Chemistry, Controversy and Conspiracy in Victorian England. Cambridge University Press, Cambridge.
08 Hamlin, C. 1990. A Science of Impurity: Water Analysis in Nineteenth Century Britain. Adam Hilger, Bristol.
09 Anon. 1881. Quarterly Journal of the Microscopical Society, October, 650.
10 Frankland, P.F. and Ward, H.M. 1892. First report to the Water Research Committee of the Royal Society, on the present state of our knowledge concerning the bacteriology of water, with special reference to the vitality of pathogenic Schizomycetes in water. Proceedings of the Royal Society of London 51, 183–280.
11 Wainwright, M. 1998. Who invented the Petri Dish? Society for General Microbiology Quarterly, August, 98–99.
12 Garner, W.E. 1948. Obituary Notice. Percy Faraday Frankland. 1858–1946. Journal of the Chemical Society 3, 1996–2005.
13 Frankland, P.F. and Frankland, P. 1884. Microorganisms in Water. Their Significance, Identification and Removal. Longmans, Green & Co., London
14 Frankland, P.F. and Ward, H.M. 1893. Second report to the Water Research Committee of the Royal Society. The vitality and virulence of *Bacillus anthracis* and its spores in potable water. Proceedings of the Royal Society of London 53, 164–317.
15 Frankland, P.F. and Ward, H.M. 1894. Third report to the Water Research Committee of the Royal Society. Part I. Further experiments on the action of light on *Bacillus anthracis*, and on the bacteria of the Thames. Part II. The behaviour of the typhoid bacillus and of the *Bacillus coli communis* in potable water. Proceedings of the Royal Society of London 56, 315–556.
16 Frankland, P.F. and Ward, H.M. 1895. Fourth report to the Water Research Committee of the Royal Society. On the biology of *Bacillus ramosus* (Fraenkel),

a Schizomycete of the River Thames. Proceedings of the Royal Society of London 58, 265–468.
17 Downes, A. and Blunt, T.P. 1877. Researches on the effect of light upon bacteria and other organisms. Proceedings of the Royal Society of London 26, 488–500.
18 Ward, H.M. 1892. Experiments on the action of light on *Bacillus anthracis*. Proceedings of the Royal Society of London 52, 393–400.
19 Ward, H.M. 1893. Further experiments on the action of light on *Bacillus anthracis*. Proceedings of the Royal Society of London 53, 23–44.
20 RS, Notes and Minutes of the Water Research Committee.

Chapter seven: Cambridge – the fatal challenge

01 RBGK, Director's correspondence, 59, 66.
02 Ward, H.M. 1890. The relations between host and parasite in certain epidemic diseases of plants. Proceedings of the Royal Society 47, 213–216.
03 RBGE, letter from Thiselton-Dyer to I Bayley-Balfour, dated 30 December 1907
04 Seward, A.C. 1932. Sir Francis Darwin. Obituary. Proceedings of the Royal Society, 110, i–v.
05 Walters, S.M. and Stowe, E.A. 2001. Darwin's Mentor. John Stevens Henslow 1796–1861. Cambridge University Press, Cambridge.
06 Walters, S.M. 1981. The Shaping of Cambridge Botany. Cambridge University Press, Cambridge.
07 Green, R.G. 1914. A History of Botany in the United Kingdom. JM Dent, London.
08 Thiselton-Dyer, W. 1907. Harry Marshall Ward. New Phytologist 6, 1–9.
09 Hill, A.W. 1942. Walter Gardiner. Obituary Notices of Fellows of the Royal Society 3, 984–1004.
10 The Cambridge Express, dated 5 March 1904.
11 Godwin, H. 1935. Cambridge and Clare. Cambridge University Press, Cambridge.
12 Thiselton-Dyer, W. 1911. Harry Marshall Ward. Proceedings of the Royal Society, B, 83, 1–14.
13 Anon. 1904. The opening of the new botanical school at Cambridge. New Phytologist 3, 61–62.
14 HU, letter dated 27 March 1897.
15 Vines, S.H. 1906. Prof. H. Marshall Ward, FRS. Nature 74, 493–495.
16 BO.
17 BO.
18 Ward, H.M. 1905. Recent researches on the parasitism of fungi. Annals of Botany 19, 1–54.
19 Ward, H.M. 1903. On the histology of *Uredo dispersa*, Erikss., and the 'Mycoplasm' hypothesis. Philosophical Transactions of the Royal Society 71, 353–354.
20 Balfour, I.B. 1906. The presidential address. Proceedings of the Royal Society 71, 218–232.
21 Whetzel, H.H. 1918. An outline of the History of Plant Pathology. WB Saunders, Philadelphia.
22 Christensen, C.M. 1984. E.C. Stakman, Statesman of Science. American Phytopathological Society, St Paul, MN.

Chapter eight: Legacies

01 Thiselton-Dyer, W. 1907. Harry Marshall Ward. New Phytologist 6, 1–9.
02 Kingdon Ward, F. 1913. The Land of the Blue Poppy. Cambridge University Press, Cambridge.
03 BL, 55515 f233.
04 BL, 55516 f266.
05 RS, Notes of the Scientific Relief Committee.
06 Lyte, C. 1989. Frank Kingdon-Ward. The Last of the Great Plant Hunters. John Murray, London.
07 Stakman, E.C. 1915. Relation between *Puccinia graminis* and plants highly resistant to its attack. Journal of Agricultural Research 4, 193–199.
08 Goodman, R.N. and Novacky, A.J. 1994. The Hypersensitive Reaction in Plants to Pathogens. American Phytopathological Society, St . Paul, MN.
09 Ward, H.M. 1905. Recent researches on the parasitism of fungi. Annals of Botany 19, 1–54.
10 Ward, H.M. 1902. Experiments on the effect of mineral starvation on the parasitism of the uredine fungus, *Puccinia dispersa*, on a species of *Bromus*. Proceedings of the Royal Society 71, 138–151.
11 RGBE, letter from Thiselton-Dyer to I Bayley-Balfour, dated 30 December 1907.
12 Ward, H.M. 1887. On the tubercular swellings on the roots of *Vicia faba*. Philosophical Transactions of the Royal Society of London, 178B, 539–562.
13 Godwin, H. 1985. Cambridge and Clare. Cambridge University Press, Cambridge.
14 Marryat, D.C.E. 1907. Notes of the infection and histology of two wheats immune to the attacks of *Puccinia glumarum*, yellow rust. Journal of Agricultural Science 2, 129–138.
15 Christensen, C.M. 1984. E.C. Stakman, Statesman of Science. American Phytopathological Society, St Paul, Minnesota.
16 Peterson, P.D. 2001. E.M. Freeman: early research on cereal diseases and the rise of plant pathology at the University of Minnesota. Annual Review of Phytopathology 39, 13–26.
17 Kolmer, J.A. 1997. Virulence dynamics and genetics of cereal rust populations in North America. Pages 139–152 in: The Gene-for-Gene Relationship in Plant-Parasitic Interactions. I.R. Crute, E.B. Holub, and J.J. Burdon, eds. CABI, Wallingford.
18 Biffen, R.H. 1907. Studies in the inheritance of disease resistance. Journal of Agricultural Science 2, 109–128.
19 Allard, R.W. 1960. Principles of Plant Breeding. John Wiley, London.
20 Green, R.G. 1914. A History of Botany in the United Kingdom. J.M. Dent, London.
21 Ramsbottom, J. 1948. Presidential address. Transactions of the British Mycological Society 30, 22–39.
22 Ainsworth, G.C. 1976. Introduction to the History of Mycology. Cambridge University Press, Cambridge.
23 Ramsbottom, J. 1948. The British Mycological Society. Transactions of the British Mycological Society 30, 1–16.
24 Smith, E.F. 1927. Fifty years of pathology. Pages 12–50 in: Proceedings of the International Congress of Plant Sciences, vol.1. J.M.Duggar, ed. George Banta, New York.
25 Large, E.C. 1940. Advance of the Fungi. Jonathan Cape, London.

26 English, M.P. 1987. Mordechai Cubitt Cooke. Biopress Ltd., Bristol.
27 RGBE, letter from Owens College to I. Bayley Balfour, dated 17 April 188?
28 RBGK, Director's correspondence, 58 65.
29 RBGK, Director's correspondence, 59 66.
30 Vines, S.H. 1906. Prof. H Marshall Ward, FRS. Nature 74, 493–495.
31 Bower, F.O. 1906. Harry Marshall Ward FRS. Journal of Botany 44, 422–425.

Appendix

Publications of Harry Marshall Ward

I. ORIGINAL RESEARCH AND REVIEWS

(A) SINGLE-AUTHOR

1880. On the embryo-sac and development of *Gymnadenia conopsea*. Quarterly Journal of Microscopical Science 20: 1–18. (Also: Report of the British Association for the Advancement of Science 1879: 375–376.)

1880. A contribution to our knowledge of the embryo-sac in Angiosperms. Journal of the Linnaean Society, Botany 17: 519–546.

1881. On the morphology of *Hemileia vastatrix* Berk. and Br. (the fungus of the Coffee Disease of Ceylon). Quarterly Journal of Microscopical Science: 1–11.

1882. Researches on the life history of *Hemileia vastatrix*, the fungus of the 'Coffee Leaf Disease'. Journal of the Linnaean Society, Botany 19: 299–335.

1882. Researches on the morphology and life-history of a tropical Pyrenomycete (*Asterina*). Quarterly Journal of Microscopical Science 22: 347–354.

1883. Observations on the Saprolegniae. Quarterly Journal of Microscopical Science 23: 272–291.

1883. On the coffee leaf disease of Ceylon. Manchester Literary and Philosophical Society Proceedings 22: 21–25.

1883. Observations on the genus *Pythium* (Pringsh.). Quarterly Journal of Microscopical Science 23: 485–515.

1883. On the structure, development and life-history of a tropical epiphyllous lichen (*Strigula coneplanata*, Fée, fide Rev. JM Crombie). Transactions of the Linnaean Society, Botany 2: 87–119.

1883. On the morphology and the development of the perithecium of *Meliola*, a genus of tropical epiphyllous fungi. Proceedings of the Royal Society of London 34: 388–390.

1884. On the morphology and the development of the perithecium of *Meliola*, a genus of tropical epiphyllous fungi.. Philosophical Transactions of the Royal Society of London 174: 583–599.

1884. On the fungus of the salmon disease (*Saprolegnia ferax*). Manchester Literary and Philosophical Society Proceedings 23: 29–32.

1884. On the sexuality of the Fungi. Quarterly Journal of Microscopical Science 24: 262–310. (also : Biological Studies of the Owens College 1: 23–63).

1885. Roots. Nature 31: 183–185.

1885. Notes on experiments as to the formation of starch in plants under the influence of the electric light. Nature 32: 563–564.

1886. The morphology and physiology of an aquatic myxomycete. Biological Studies of the Owens College 1: 64–86.
1886. Roots. Nature 34: 524–528.
1887. Illustrations of the structure and life-history of *Phytophthora infestans*, the fungus causing the potato disease. Quarterly Journal of Microscopical Science 27: 413–425.
1887. On the structure and life-history of *Entyloma ranunculi* (Bonorden). Philosophical Transactions of the Royal Society of London 178, B: 539–562.
1887. The tubercular swellings of the roots of the Leguminoseae. (Preliminary note.) Proceedings of the Royal Society of London 42: 331.
1888. On the tubercular swellings on the roots of *Vicia faba*. Philosophical Transactions of the Royal Society of London 178, B: 539–562.
1888. Anton De Bary. Nature, 26 January: 297–299.
1888. Some recent publications bearing on the question of the sources of nitrogen in plants. Annals of Botany 1: 325–357.
1888. A lily disease. Annals of Botany 2: 319–378.
1888. Illustrations of the structure and life-history of *Puccinia graminis*, the fungus causing the Rust of wheat. Annals of Botany 2: 217–222.
1888. Timber and some of its diseases. Nature 37: 182–186 (continued), 204–207, 227–229, 251–254, 275–279 and 514–516.
1888. Timber and some of its diseases. Nature 38: 108–111 (continued), 127–130, 270–272, 297–299, 367–370.
1889. Smut fungi: recent discoveries as to the nature and action of Ustilagineae. Gardener's Chronicle 5: 233–235.
1889. Chlorosis. Gardener's Chronicle 5: 462–463.
1890. On the tubercles on the roots of leguminous plants, with special reference to the Pea and Bean. Proceedings of the Royal Society of London 46: 431–443.
1890. The fungi of fermenting vessels and vats. The Brewers' Journal, 33–36 (continued), 116–118, 223–226, 408–412, 524–527, and 623–627, also (1891) 40–42.
1890. The relations between host and parasite in certain epidemic diseases of plants: The Croonian Lecture. Philosophical Transactions of the Royal Society of London 47: 213–216.
1890. The teaching of botany. Nature 42: 620–626.
1891. Notes on botanical characters serving to distinguish the principal British forest trees. In: Manual of Forestry, Vol.II. (Appendix). W. Schlich. Bradbury, Agnew & Co , London, UK.
1892. The Ginger-beer Plant, and the organisms composing it: a contribution to the study of fermentation – yeasts and bacteria. Philosophical Transactions of the Royal Society of London, Series B, 183: 125–197.
1892. The diseases of conifers. Journal of the Royal Horticultural Society 14: 124–150.
1892. Symbiosis and symbiotic fermentations. Transactions of the Brewing Industry 5: 55–77.
1892. On the characteristics or marks employed for classifying the Schizomycetes. Annals of Botany 6: 103–144.
1892. Experiments on the action of light on *Bacillus anthracis*. Proceedings of the Royal Society of London 52: 393–400.
1893. Further experiments on the action of light on *Bacillus anthracis*. Proceedings of the Royal Society of London 53: 23–44.

1893. Experimental investigations on the behaviour of *Bacillus anthracis* in water. Proceedings of the Royal Society of London 53: 245–312.
1894. The action of light on bacteria III. Philosophical Transactions of the Royal Society 185, B: 961–986.
1894. The action of light on bacteria and fungi. Photographic Journal 18: 270–278. (Also: Proceedings of the Royal Institution 14: 259–279).
1894. Further experiments on the action of light on *Bacillus anthracis*, and on the bacteria of the Thames. Proceedings of the Royal Society of London 56: 315–394.
1894. Recent investigation and ideas on the fixation of nitrogen by plants. Nature 49: 511–514.
1895. New aspects of an old agricultural question. Science Progress, 251.
1895. The formation of bacterial colonies. Annals of Botany 9: 653–657, (also: Report of the British Association for the Advancement of Science).
1895. On the biology of *Bacillus ramosus* (*Fraenkel*), a schizomycete of the River Thames. Proceedings of the Royal Society of London 58: 265–468.
1895. A false bacterium. Annals of Botany 9: 657–658. (Also: Report of the British Association for the Advancement of Science, 850).
1895. The action of light on bacteria. Cambridge Philosophical Society Proceedings 8: 128–129.
1897. Fifth report to the Water Research Committee of the Royal Society. Proceedings of the Royal Society of London 61: 415 –423.
1897. On *Peziza aurantia*. Annals of Botany 11: 339–341.
1897. On the Ginger-beer Plant. Annals of Botany 11: 341–344.
1897. Microscopic gardening. Journal of the Royal Horticultural Society 21: 4–24.
1898. On the biology of *Stereum hirsutum* (*Fr.*). Philosophical Transactions of the Royal Society 189, B: 123–134.
1898. A violet bacillus from the Thames. Annals of Botany 12: 59–74.
1898. Some Thames bacteria. Annals of Botany 12: 287–322.
1898. Some brewing botanical problems. Journal of the Federated Institutes of Brewing 4: 334–363.
1899. Some Thames bacteria. Annals of Botany 13: 197–251.
1899. *Onygena equina*, Willd., a horn-destroying fungus. Philosophical Transactions of the Royal Society 191, B:269–291.
1899. Thames bacteria III. Annals of Botany 13: 197–251.
1899. The potency and prepotency of pollen. Nature 61: 470–471.
1899. Symbiosis. Annals of Botany 13: 549–562. (Also: Report of the British Association for the Advancement of Science 1899, 692–697).
1899. Some methods for use in the culture of algae. Annals of Botany 13: 563–566.
1901. Notes on some of the rarer or more interesting fungi collected during the past year. Cambridge, Proceedings of the Philosophical Society 11: 93–100.
1901. The nutrition of fungi. Presidential address. Transactions of the British Mycological Society 1: 124–142.
1901. On the biology of a *Naemetelia*, Fr. Transactions of the British Mycological Society 1: 143–150.
1901. The bromes and their rust fungus, *Puccinia dispersa* (Erikss. et Henn.). Annals of Botany 15: 560–562.
1902. On the relations between host and parasite in the bromes (*Bromus*) and their brown rust, *Puccinia dispersa* Erikss. Annals of Botany 16: 233–315.

1902. On pure cultures of a uredine, *Puccinia dispersa*, Erikss. Proceedings of the Royal Society of London 69: 451–466.
1902. On the question of 'predisposition' and 'immunity' in plants. Cambridge, Proceedings of the Philosophical Society 11: 307–328.
1902. Experiments on the effects of mineral starvation on the parasitism of the uredine fungus, *Puccinia dispersa*, on species of *Bromus*. Proceedings of the Royal Society of London 71: 138–151.
1903. On the histology of *Uredo dispersa*, Erikss., and the 'Mycoplasm' Hypothesis. Proceedings of the Royal Society of London 71, 218–232.
1905. Recent researches on the parasitism of fungi. Annals of Botany 19: 1–54.

HMW also contributed notes to the Encyclopaedia Britannica: Schizomycetes (1885 edtn); Bacteriology; Fungi; and Pathology of Plants (1902 edtn).

(B) JOINT-AUTHOR

1887. Ward, H. M and Dunlop, J. On some points in the histology and physiology of the fruits and seeds of *Rhamnus*. Annals of Botany 1: 1–26.
1892. Frankland, P. F. and Ward, H. M. First report to the Water Research Committee of the Royal Society, on the present state of our knowledge concerning the bacteriology of water, with special reference to the vitality of pathogenic Schizomycetes in water. Proceedings of the Royal Society of London 51: 183–280.
1893. Frankland, P. F. and Ward, H. M. Second report to the Water Research Committee of the Royal Society. The vitality and virulence of *Bacillus anthracis* and its spores in potable water. Proceedings of the Royal Society of London 53: 164–317.
1894. Frankland, P. F. and Ward, H. M. Third report to the Water Research Committee of the Royal Society. Further experiments on the action of light on *Bacillus anthracis*, and on the bacteria of the Thames. Proceedings of the Royal Society of London 56: 315–556.
1895. Frankland, P. F. and Ward, H. M. Fourth report to the Water Research Committee of the Royal Society. On the biology of *Bacillus ramosus* (Fraenkel), a Schizomycete of the River Thames. Proceedings of the Royal Society of London 58: 265 –468.
1898. Ward, H. M. and Dale, J. On *Craterostigma pumilum*, Hochst., a rare plant from Somaliland. Journalof the Linnaean Society, Botany 5: 343–355.
1899. Ward, H. M. and Green, J. R. A sugar bacterium. Proceedings of the Royal Society of London 65: 65–84.
1902. Freeman, E. M. and Ward, H. M. The seed-fungus of *Lolium temulentum* L., the Darnel. Proceedings of the Royal Society of London 71: 27–30.

II. BOOKS (of which HMW was the author, unless stated otherwise)

1887. Sachs, J. von, Lectures on the Physiology of Plants. (Trs. by H.M. Ward). Clarendon Press, Oxford, UK.
1889. Timber and Some of its Diseases. Macmillan & Co., London, UK.
1891. Diseases of Plants. Society for the Promoting of Christian Knowledge, London, UK.

1892. The Oak, a Popular Introduction to Forest Botany. Kegan Paul & Co., London, UK.
1894. Laslett, T., Timber and Timber Trees: Native and Foreign, 2nd edn. Revised by H. M. Ward. MacMillan & Co, London, UK.
1894. Hartig, R., Textbook of the Diseases of Trees. Trs W. Somerville. Revised and ed. H. M. Ward. Macmillan & Co., London, UK.
1901. Disease: In Plants. Macmillan & Co., London, UK.
1901. Grasses: A Handbook for Use in the Field and Laboratory. Cambridge University Press, Cambridge, UK.
1904. A Handbook of Forest Botany for the Woodlands and Laboratory, Vol. I: Buds and Twigs. Cambridge University Press, Cambridge, UK.
1904. A Handbook of Forest Botany for the Woodlands and Laboratory,Vol. II: Leaves. Cambridge University Press, Cambridge, UK.
1905. A Handbook of Forest Botany for the Woodlands and Laboratory,Vol. III: Flowers and Inflorescences. Cambridge University Press, Cambridge, UK.

III. REPORTS NOT INCLUDED ABOVE

1880a. Coffee Leaf Disease. Preliminary Report by the Government Cryptogamist. Colombo Sessional Papers. XVII: 141–144.
1880b. Coffee Leaf Disease. Second Report by H. Marshall Ward. Colombo Sessional Papers. L: 555–566.
1881. Coffee Leaf Disease. Third Report by H. Marshall Ward. Colombo Sessional Paper. XVII: 227–262.
1891. On a simple apparatus for the cultivation of small organisms in hanging drops, and in various gases, under the microscope. Report of the British Association for the Advancement of Science: 678–679.
1897. The advance in our knowledge of the Fungi since 1860. Report of the British Association for the Advancement of Science, 831–859.
1898. A potato disease. Report of the British Association for the Advancement of Science, 1046–1048.
1898. *Penicillium* as a wood-destroying fungus. Report of the British Association for the Advancement of Science, 1048–1049.
1900. Embryonic tissues. Report of the British Association for the Advancement of Science, 943–944.
1902. The bromes and their brown rust (*Puccinia dispersa,* Erikss. et Henn.). Report of the British Association for the Advancement of Science.

These lists were compiled from :

1896. Anonymous. Catalogue of Scientific Papers of Fellows of the Royal Society of London (1874–1883). XI. London: CJ Clay & Sons.
1906. Balfour, I.B. Presidential Address. Transactions and Proceedings of the Botanical Society of Edinburgh. 23: 218–232.
1925. Anonymous. Catalogue of Scientific Papers of Fellows of the Royal Society of London, Fourth Series (1884–1902). XIX. Cambridge: Cambridge University Press.

Author's Acknowledgements

My sincere thanks are due to a host of men and women who helped me write this biography. In particular, I thank Harry's great-grandson, Oliver Tooley Watkins, for allowing me to read and quote from Harry's letters to his family. I thank too those local historians who helped me piece together parts of Harry's early life, Ifor Barton (Lincoln), Eleanor Harris (Hereford), Sheila Jewell (Exeter), Philip Jones (Nottingham) and Judith Moore (Manchester).

The librarians and archivists whom I encountered were unfailingly helpful. They include Clara Anderson, Martin Carr and Christine Woollett (Royal Society of London), Nicholas Bennett (Lincoln Cathedral Library), Michael Bott (University of Reading), David Brady and James Peters (University of Manchester), Stella Brecknell and Colin Harris (University of Oxford), Lisa Decesare (Harvard University), Gina Douglas (Linnean Society), Barbara Gent (Giggleswick School), Paula Gerrard (Brunel University), Claire Glazebrook and Candace Guite (Christ's College, Cambridge), Graham Hardy (RBG, Edinburgh), Chris Jakes (Cambridgeshire Library), Elizabeth James (British Library), Brian Jenkins and Richard Savage (University of Cambridge), Jim Kay, Michelle Losse and Kate Pickard (The Herbarium, RBG, Kew), Nalini Kekulandera (University of Peradeniya), Stella Lowe (Manchester Lit. & Phil. Soc.), Simon May (St. Paul's School), Hilary McEwan (Imperial College, London), Colin Shaw (Sutton Valence School), Angela Todd (Carnegie Mellon University, Pittsburgh), Frances Ward (Greenwich Library), and Joan Williams (Hereford Cathedral Library).

Others who provided vital information about, or interpretations of, Harry's life were Raymond Anderson (Burton on Trent), Don Boney (Yelverton), Gerhart Drews (University of Freiburg), Peter Grubb, Anne Stowe and Max Walters (University of Cambridge), Rod Hunt (University of Sheffield), Charles Lyte (Exeter), Richard Morris (Brunel University), Charles Nelson (Wisbech), Emma Reisz and Ronald Wood (Imperial College), and Gail Vines (Lewes).

I thank those who bravely read and advised upon draft chapters, including Rory Akam (Stowe School), Juliet Frankland (ITE, Lancaster), Peter Harmon, Keith Jones, Terry Mansfield and Nigel Paul (Lancaster University), and Janet Philpott (High Wycombe Royal Grammar School).

For his interest throughout the project and for encouraging me to propose to APS that it should publish this book, I thank Paul Peterson (Clemson University). For his sympathetic and constructive editing of the text, I thank

Gareth Hughes (University of Edinburgh). The APS production team in St. Paul, MN was most helpful throughout.

Finally, and mostly, I thank my wife, Mary. She not only transcribed letters, organised references, read drafts, and helped with a multitude of practical tasks, but encouraged, supported and stimulated my interest in Harry Marshall Ward – each at exactly the right time.

Index